Estimation of Uncertainty of Wind Energy Predictions

Maritime Logistik

Herausgegeben von
Frank Arendt und Burkhard Lemper

Band 9

Zur Qualitätssicherung und Peer Review der vorliegenden Publikation

Die Qualität der in dieser Reihe erscheinenden Arbeiten wird vor der Publikation durch Herausgeber der Reihe geprüft.

Notes on the quality assurance and peer review of this publication

Prior to publication, the quality of the work published in this series is reviewed by editors of the series.

David Zastrau

Estimation of Uncertainty of Wind Energy Predictions
With Application to Weather Routing and Wind Power Generation

Bibliographic Information published by the Deutsche Nationalbibliothek
The Deutsche Nationalbibliothek lists this publication in the Deutsche Nationalbibliografie;
detailed bibliographic data is available in the internet at http://dnb.d-nb.de.

Zugl.: Bremen, Univ., Diss., 2016

Library of Congress Cataloging-in-Publication Data
Names: Zastrau, David, author.
Title: Estimation of uncertainty of wind energy predictions : with
application to weather routing and wind power generation / David Zastrau.
Description: Frankfurt am Main : Peter Lang GmbH, 2017. | Series: Maritime
logistics, ISSN 1868-369X ; vol. 9 | Includes bibliographical references and index.
Identifiers: LCCN 2017002422| ISBN 9783631718858 | ISBN 9783631718957 (E-PDF)
| ISBN 9783631718964 (EPUB) | ISBN 9783631718971 (MOBI)
Subjects: LCSH: Wind power. | Winds--Measurement. | Estimation theory.
Classification: LCC TJ820 .Z37 2017 | DDC 621.31/2136--dc23 LC record available at
https://lccn.loc.gov/2017002422

D 46
ISSN 1868-369X
ISBN 978-3-631-71885-8 (Print)
E-ISBN 978-3-631-71895-7 (E-PDF)
E-ISBN 978-3-631-71896-4 (EPUB)
E-ISBN 978-3-631-71897-1 (MOBI)
DOI 10.3726/b11013

© Peter Lang GmbH
Internationaler Verlag der Wissenschaften
Frankfurt am Main 2017
All rights reserved.

PL Academic Research is an Imprint of Peter Lang GmbH.

Peter Lang – Frankfurt am Main · Bern · Bruxelles · New York ·
Oxford · Warszawa · Wien

All parts of this publication are protected by copyright. Any
utilisation outside the strict limits of the copyright law, without
the permission of the publisher, is forbidden and liable to
prosecution. This applies in particular to reproductions,
translations, microfilming, and storage and processing in
electronic retrieval systems.

This publication has been peer reviewed.

www.peterlang.com

Contents

Glossary IX

Acronyms XI

Acknowledgements XIII

1. Introduction 1
   1.1. Uncertainty in wind energy predictions 1
   1.2. Approaches from the literature 2

2. Uncertainty in Wind Power Generation and
   Weather Routing 5
   2.1. Weather prediction uncertainty in wind power generation (WPG) 6
      2.1.1. Offshore wind power logistics 6
   2.2. Prediction uncertainty in weather routing 7
      2.2.1. Wind propulsion systems (WPS) 9
      2.2.2. Speed power curve 15
      2.2.3. Wind resistance 16
      2.2.4. Wave resistance 17
      2.2.5. Ship propulsion energy 18
      2.2.6. Prediction uncertainty with WPS 20
   2.3. Wind power generation versus weather routing 20
   2.4. Weather forecasting 21
      2.4.1. Numerical weather analyses and predictions 22
      2.4.2. Limitations and trends in weather forecasting 24

3. Statistical Patterns of Uncertainty in
   Weather Predictions 29
   3.1. Prediction error metrics 29

3.2. Predictions for the North Atlantic Ocean 30
    3.2.1. DWD wind and wave predictions 31
    3.2.2. Regional and seasonal prediction uncertainty 33
3.3. Predictions for the North and Baltic Seas 34
    3.3.1. FINO measurements 34
    3.3.2. Prediction uncertainty 37
    3.3.3. Wave prediction uncertainty 37
    3.3.4. Wind prediction uncertainty 42
    3.3.5. Conclusions 47

4. Estimation of Prediction Uncertainty 49
4.1. Theoretical and empirical models 49
4.2. Ensemble prediction systems 50
    4.2.1. Multi-model and multi-analyses ensembles 50
    4.2.2. Super ensembles 51
4.3. Statistical methods 51
    4.3.1. Probability density estimation methods 52
    4.3.2. Clustering 53
    4.3.3. Prediction intervals 54
    4.3.4. Quantile regression 55

5. The Quantile Regression Model 57
5.1. Linear quantile regression (QR) 57
5.2. Regressors for the QR model 59
    5.2.1. Principal component analysis 60
    5.2.2. North Atlantic Oscillation Index (NAOI) 61
    5.2.3. Other climatological indices 62
    5.2.4. Statistical moments 63
    5.2.5. Local Energy Distribution Moments (LEDM) 65
    5.2.6. Relation between LEDM and NAOI 69

6. Implementation 71
6.1. Database with historical weather predictions................................71
6.2. A* route optimization 72
6.3. Route optimization with uncertainty 75
6.4. Quantile regression 76

7. Evaluation and Results..79
7.1. Evaluation of prediction intervals..79
7.2. Application to wind-assisted sailing propulsion 80
7.3. Prediction intervals for ship propulsion energy.............................83
7.4. Prediction intervals for travel time 86
7.5. Uncertainty in weather routing with WaSP...................................88
7.6. Prediction interval benchmarks..89
7.7. Runtime discussion 92
7.8. Annual cost savings for a multi-purpose carrier............................92

8. Conclusions and Outlook..95
8.1. New approach and scientific contribution.....................................95
8.2. Outlook...96
8.2.1. Prediction uncertainty of wind turbine power output................96
8.2.2. Uncertainty in weather routing...96
8.2.3. Further applications 97

Bibliography..99

List of Figures 107

List of Tables... 111

A. Parameters of Wind Propulsion Systems 113

B. Parameters of the wind and wave resistance models 115

C. UML diagrams of the Java implementation 117

D. The Global Sea Model (GSM) 123

Glossary

E_A actual energy (calculated with weather analyses) 18
E_P predicted energy (calculated with weather prediction) 18
E_T estimated propulsion energy in A* for the complete route 72
$E_{P,Err}$ prediction error for E_A 18
FN normalized force of WPS 10
F_{WPS} force of WPS 10
I^α prediction interval 54
I_R^α reliability of prediction intervals (coverage) 54
I_S^α skill score of prediction intervals (uncertainty & coverage) 55
I_U^α sharpness of prediction intervals (uncertainty) 54
I_{Err}^α error of prediction intervals 54
P_{WPS} power of WPS 10
P_{Wa} power of waves 17
P_{Wi} power of wind 17
P_{cw} propulsion power of the ship in calm water 15
Q quantile 57
QR_{LEDM} quantile regression model with LEDM regressors 69
T draft of ship 15
V_s speed of ship 15
α significance coefficient of prediction intervals 52
β weight vector for quantile regression model 58
η_T efficiency of transformation from motor power to propulsion power 10
ρ_α check function 58
k size of **roi** 67
roi region of interest (a vector with coordinates) 68
x vector with regressors for the quantile regression model 57

dynamic programming mathematical technique to solve optimization problems by constructing a solution from multiple partial solutions 8

ensemble prediction set of numerical weather predictions (NWP) calculated with slightly different initial model conditions 50

genetic algorithm search heuristic which imitates processes from natural selection to find the solution to a problem 8

great circle circle on the earth sphere with maximum perimeter 9

jack-up vessel ship which has been designed for the footing and construction of offshore wind power plants 6

lagged ensemble consists of multiple succeeding weather forecasts which predict the same future state of weather 50

orthodrome part of the great circle and the shortest connection between two points A and B on the earth sphere (equations are described in (Hoschek, 1984)) 72

parametric rolling dangerous oscillatory roll motion of a ship with moderate or large amplitude mostly caused by waves at right angles to the ship 8

Parzen window density estimation interpolates a between values in a data sample to estimate the probability density function from which the sample was derived 52

percentile defines the value which is greater than i % (i-th percentile) of the values of a dataset 57

poor man's ensemble set of forecasts from different weather forecasting models which predict the same future state of weather 50

quantile defines the value which is greater than $i \cdot 25$ % (i-th quantile) of the values of a dataset 57

skill score quality measure for prediction intervals that takes into account interval sharpness and reliability 55

speed power curve is a function that shows ship propulsion energy for desired ship speeds 15

WaMoS II wave radar developed by the OceanWaves GmbH 37

Acronyms

BIAS mean signed error 30

DWD German Weather Service 22

EPS ensemble prediction system 51

FINO research platforms in the North and Baltic Seas 34

LEDM Local Energy Distribution Moments 68

MAE mean absolute error 29
MDPS mean direction of primary swell 32
MDWI mean wind direction at 10 m above sea level 42
MDWW mean direction of wind waves 32
MU10 mean wind speed at 10 m above sea level 42

NAOI North Atlantic Oscillation index 59
NWP numerical weather prediction 22

PCA principal component analysis 60

QR quantile regression 55

RMSE root-mean-square error 30

SWH significant wave height 32

TP mean wave period 32

UML Unified Modelling Language 19

WaSP wind-assisted ship propulsion 5
WEP wind energy prediction 1
WPS wind propulsion system 1

Acknowledgements

I am very grateful for the guidance and support by Prof. Otthein Herzog and Prof. Michael Schlaak. I would also like to thank Prof. Elsner who initiated the ROBUST project which was the framework for this thesis. I am grateful to all the associates in Bremen and in Emden for helping me with fruitful discussions or just for being great colleagues. Furthermore I would like to thank Thorsten Bielefeld, Roland Grashorn, Andreas Traumüller and Heiko Stein whose bachelor theses contributed to the implementation of this work.

I am grateful to Dr. Thomas Bruns for his consultation and for providing historical weather predictions by the German Weather Service which where essential to this work.

And I would like to thank the "VolkswagenStiftung" for financing my thesis.

David Zastrau
Bremen, Germany, January 2016

Dedicated to my family and my friends

1. Introduction

Abstract: *This chapter gives a short introduction to the problem of uncertainty in wind energy predictions (WEPs) and describes the new approach in this work to estimate WEP uncertainty.*

Contents

1.1. Uncertainty in wind energy predictions 1
1.2. Approaches from the literature 2

1.1. Uncertainty in wind energy predictions

Since fossil energy resources are decreasing, efforts are made to replace them with regenerative energy resources. Examples are wind propulsion systems (WPSs) and wind energy generation. Yet so far there exist only few ships with WPS and the proportion of renewable energy sources in the European Union is only 8.5% (Klessmann *et al.*, 2011). One of the reasons for this is weather-induced wind energy variability as shown in fig. 1.1.

Figure 1.1: Aggregate electricity demand in Denmark (West) vs. total hourly wind production. "1" indicates first hour of the day. Image & text: (Kunz et al., 2014, p. 161).

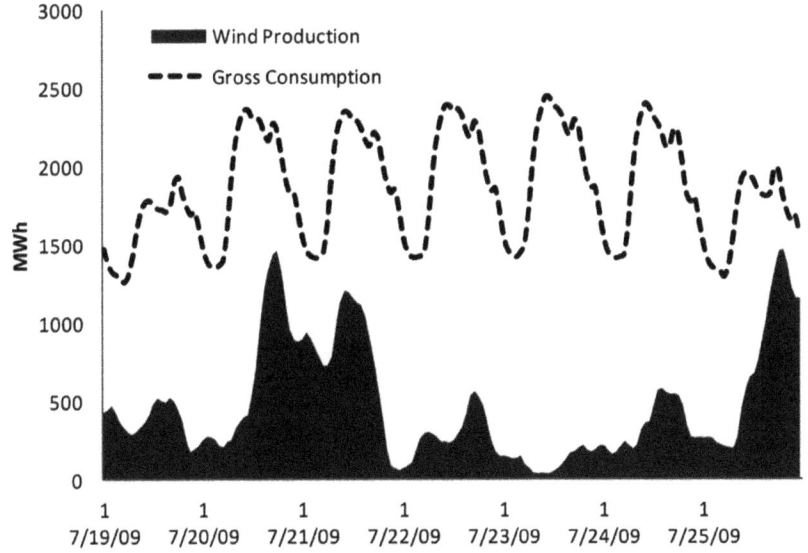

The aggregate electricity demand in western Denmark follows a regular periodic pattern whereas the wind production varies significantly. If the numerical weather prediction (NWP) is inaccurate, these variations are unpredictable. Yet if the NWP is accurate the wind energy prediction (WEP) is also accurate and the wind energy can be used more efficiently:

- In wind energy generation the electricity demand can be adjusted to the available wind production, e.g., the industrial production can be adjusted to the available wind electricity production.
- In offshore logistics, the shipping routes can be optimized (*weather routing*) to maximize the efficiency of the WPS.

Therefore, it is important to estimate the uncertainty of WEP in order to use the wind energy efficiently. The uncertainty of WEP depends on NWP accuracy.

1.2. Approaches from the literature

The accuracy of numerical weather predictions varies depending on the general weather conditions. Therefore three different principal approaches exist to estimate the uncertainty in NWP:

- Ensemble Prediction Systems (EPS) calculate a set of possible NWP thus showing a spread of possible weather scenarios (NWP uncertainty). For this purpose the NWP is calculated multiple times with a different model initialization or parameters. The literature on this subject is reviewed in sec. 4.2.
- Clustering, principal component analysis and kernel density estimation can identify *typical* weather categories in historical data and estimate NWP uncertainty of each weather category. A new weather situation is assigned to a category and to the associated NWP uncertainty. A brief literature review on this topics is given in secs. 4.3.1–4.3.2.
- NWP uncertainty can also be shown by the width of prediction intervals. These are calculated based on historical prediction uncertainty with quantile regression (QR) based on a set of custom variables (regressors) that indicate weather variability, e.g., mean local wind speeds. The literature is reviewed in secs. 4.3.3–4.3.4.

The NWP uncertainty which is calculated by these three methods is used to calculate WEP uncertainty as outlined by eq. 1.1.

$$\text{NWP uncertainty} \xrightarrow{\text{non-linear model}} \text{WEP uncertainty} \quad (1.1)$$

In this approach the relation between NWP and WEP uncertainties is modeled explicitly by a non-linear model, e.g., by calculating the propulsion energy of a WPS for different predicted wind speed and direction values (e.g., the output of an EPS). There is no general model for the relation between NWP uncertainty and WEP uncertainty (except for probabilistic approaches like Monte Carlo Sampling which assume that the probability distributions of weather parameters are known).

In this work two variables, called *Local Energy Distribution Moments* (**LEDM**), are introduced which summarize the uncertainty in the WEP in a region of interest. Since LEDM are simply the mean and variance of the WEP in the region of interest the local energy distribution is described without making strong distributional assumptions. With **LEDM** the standard linear quantile regression model can be applied to calculate WEP uncertainty (eq. 1.2). The non-linear relation in eq. 1.1 does not need to be modeled explicitly here. Thus WEP uncertainty can be calculated even for models with complex relations between NWP uncertainty and WEP uncertainty.

$$\text{NWP} \rightarrow \text{LEDM} \xrightarrow{\text{linear model}} \text{WEP uncertainty} \qquad (1.2)$$

In this work the uncertainty, i.e., a prediction interval for a specified level of confidence, in WEP is calculated with linear quantile regression with LEDM regressors (QR-LEDM). The limits are also bias-corrected, which means that the WEP does not necessarily equals the center of the prediction interval. This is because QR-LEDM does not estimate a variance (or a distribution) but two quantiles instead which are independent from each other and together constitute a prediction interval. QR-LEDM takes advantage of the non-parametric and efficient QR approach but also takes into account different possible weather scenarios in the region of interest.

Since the method solves a linear optimization problem instead of estimating the probability density function of an unknown distribution it is not necessary to estimate a bandwidth parameter or a clustering parameter. The resulting prediction intervals yield higher reliability, sharpness and skill scores compared with simple QR prediction intervals. In other words the average width of prediction intervals is smaller and yet a higher percentage of actual energy values lies within the limits of the prediction intervals.

Furthermore, the number of large prediction errors decreases significantly even for predictions with 7 days look-ahead time. This shows that LEDM contain relevant information about the uncertainty in the WEP. QR-LEDM can be used to estimate uncertainty of different types of weather-dependent energy predictions (e.g., ship propulsion energy or wind power generation). If a set of historical WEP predictions is available, QR-LEDM is very efficient since it only involves solving a linear optimization problem. This is much faster than ensemble forecasting. QR-LEDM is integrated here into a weather routing application for the optimization of the propulsion energy of a ship with a WPS where it allows to skip weather routing if prediction uncertainty is high.

As a consequence of this, the average prediction error for the ship propulsion energy is reduced by 42 %, when applied over a year to a special ship route (Europe-USA). This shows that WEP uncertainty estimates improve the predictability of the results of weather routing. Another finding is that the application of a WPS drastically increases WEP uncertainty.

2. Uncertainty in Wind Power Generation and Weather Routing

Abstract: *This chapter gives an overview about prediction uncertainty in two offshore applications: ship weather routing and offshore wind power generation. Wind-assisted ship propulsion (WaSP) is discussed in this context. The importance of weather prediction uncertainty in both applications is described. An overview is given about limitations and trends in weather forecasting. Analogue weather forecasting is identified as a general approach to estimate uncertainty of predictions based on prediction uncertainty of historical weather predictions.*

Contents

2.1. **Weather prediction uncertainty in wind power generation (WPG)** 6
    2.1.1. Offshore wind power logistics 6
2.2. **Prediction uncertainty in weather routing** 7
    2.2.1. Wind propulsion systems (WPS) 9
    2.2.2. Speed power curve 15
    2.2.3. Wind resistance 16
    2.2.4. Wave resistance... 17
    2.2.5. Ship propulsion energy... 18
    2.2.6. Prediction uncertainty with WPS 20
2.3. **Wind power generation versus weather routing**............................ 20
2.4. **Weather forecasting** 21
    2.4.1. Numerical weather analyses and predictions 22
    2.4.2. Limitations and trends in weather forecasting..................... 24

Section 2.1 describes weather prediction uncertainty in wind power generation. Section 2.2 describes prediction uncertainty in weather routing and a model to calculate the ship propulsion energy for wind-assisted ship propulsion (WaSP) systems. A WaSP is a propulsion system that combines a conventional combustion engine with an additional wind propulsion system (WPS). Section 2.3 compares wind power generation and weather routing. Section 2.4 describes basic causes of numerical weather prediction (NWP) uncertainty and describes NWP by the German Weather Service (DWD) which are used in this work.

2.1. Weather prediction uncertainty in wind power generation (WPG)

Currently, only little infrastructure exists to buffer wind energy in order to improve the robustness of the power supply. Therefore, uncertainty estimates for offshore wind power generation (WPG) are necessary to adjust the (industrial) energy consumption to the available wind energy (see fig. 1.1). And NWP uncertainty estimates are also relevant for the construction and maintenance of offshore wind power plants since many operations are only feasible during calm weather conditions.

2.1.1. Offshore wind power logistics

In 2014 the growing offshore wind industry had installed 146 offshore wind turbines with a total power of 628 MW and is currently installing another 570 plants with a total power of 2300 MW in the North and Baltic Seas (Schlalos et al., 2014). The so called jack-up vessels require a meteorological time window for the installation of a wind turbine. If the time window is missed, additional charter costs incur of often more than 100.000 € per day (Schütt & Lange, 2014). The farther the wind park is away from the coast, the broader the time window needs to be.

For example, it takes a ship two hours from the nearest harbor to reach the wind park Alpha Ventus in the German Sea (Schlalos et al., 2014) and the "crew transport ships" for maintenance operations are only operational with wave heights of up to 1.5 m (Rinne & Haasis, 2014). If a maintenance operation is urgent and the weather does not allow for ship operation than a helicopter can be used to reach the wind plant. Helicopters can safely approach wind platforms if wind speeds do not exceed 20 $\frac{m}{s}$ (Böttcher, 2013, p. 450).

But not only wind speed and wave height set meteorological boundaries for offshore logistics. Additionally, water levels and sea current can delay operations. For example, underwater maintenance by industrial divers or by remotely operated underwater vehicles (ROV) is not feasible if the sea current is too strong. Planning in offshore logistics is often optimized by simulation because the multiplicity of possible scenarios does not allow for analytical solutions (Görges et al., 2014). Established tools for planning and control are often not designed for environments which are as unpredictable as in the offshore wind industry (Scholz-Reiter et al., 2011). The objective is to develop proactive plans which take into account certain "if-else" options right from the beginning. Therefore, it is recommendable to reserve appropriate resources. Thus a complete re-planning with severe consequences for the whole supply chain can be avoided. Task-forces have been proposed to interact if a process is delayed or interrupted (Heidmann, 2014). However, the number of task forces could be reduced if their probable areas of operation would be known earlier. In addition, it would allow for a better prioritization of tasks. This would require modeling the expected uncertainty in processes with respect to the relevant boundary conditions. To map uncertainty in different planning steps accordingly it has been proposed to model the outcome of critical processes by probability distributions (Greiner & Joschko, 2014). The physical life time of mechanical components can be described by a Weibull distribution and human-induced uncertainty by a normal distribution.

Aside from material deficiencies, traffic disturbances and missing authorizations, weather is the primal source of uncertainty. Weather-related delays imply drastic additional costs and re-planning. Logistical resources such as jack-up vessels or terminals are only available for a limited time slot. Weather-related delays imply significant additional costs in offshore wind logistics which is why it has been recommended to model weather as a stochastic component. Hence, proactive planning can react in time and reallocate resources accordingly. The "Business Process Model and Notation" (BPMN) explicitly models weather not as a discrete variable but as a stochastic component. The idea is to model individual processes (e.g., weather) stochastically in order to detect critical bottlenecks because modeling all the main processes and their possible outcomes is impossible.

Since the relation between weather and the variable of interest is often non-linear, it is recommendable to calculate the statistical probability distribution directly for the variable of interest. For example, a probability distribution could show the probability that a maintenance operation has to be canceled instead of multiple probability distributions which show the probabilities of separate weather parameters (wind speed, wave height, etc.). A similar case for the routing of ships is discussed in-depth in the next section.

2.2. Prediction uncertainty in weather routing

Weather routing summarizes methods which adjust a scheduled shipping route based on the latest weather forecast in order to minimize the ship propulsion energy. Estimating the uncertainty of predictions enables to avoid routes for which only uncertain predictions exist. Weather routing can offer fuel savings especially for long routes—if waters are navigable without restrictions. A study by the International Maritime Organization (IMO) (MEPC58/INF.21, 2008) estimated that weather routing for modern ships can result in fuel savings between 2–4%. WaSP potentially increases these numbers essentially.

> *Traditionally,* weather routing was based on a set of ship performance curves, an example is depicted in (Bowditch, 2011, pp. 542–547). These curves define ship speeds for given wave heights and wave directions and for a fixed ship engine power in terms of revolutions per minute. This methodology only adjusts the ship heading and focuses on *storm avoidance,* it does not permit to implement strategies that are based on variable ship engine power, e.g., riding out the weather and seas.

Based on technical and scientific innovations, the traditional weather routing approach has been extended to achieve a more realistic simulation of the ship navigation. NWP uncertainty in weather routing of (sail-assisted) motor vessels is already discussed in (Hagiwara & Spaans, 1987). Some measures to reduce or estimate NWP uncertainty in weather routing are listed below.

1. Methods to improve the quality of weather predictions (e.g., super ensembles in section 4.2) for the sea and swells generated by tropical cyclones.
2. Ensemble-based methods, e.g., the safe operation envelope by (Spaans & Stoter, 1995), are used to avoid dangerous scenarios (e.g., parametric rolling).
3. Selective climatology can improve weather routing in El Niño and La Niña phases or during extreme weather events (e.g., hurricanes) which can interrupt ocean currents for several days.
4. New algorithms (e.g., based on dynamic programming) take into account uncertainty of weather predictions as probabilistic variables in the model.

The optimization problem. Different algorithms and concepts have been proposed to improve weather routing compared with the traditional isochrones method by (James, 1957). For example, (Tsujimoto & Tanizawa, 2006) optimize the route and the number of propeller revolutions with the augmented Lagrange multiplier method. (Shao *et al.,* 2012) use dynamic programming to optimize the ship's heading and the number of propeller revolutions. Uncertainty in NWP is an important issue, because the travel time can extend to multiple weeks. The objective of weather routing is to find an optimum between soft constraints (fuel consumption, crew comfort) and hard constraints (safety of crew and cargo, travel time limit). Constraints can be contradictory, solutions are to either specify weights for the different constraints or to set fixed thresholds for all but one constraint.

> An important constraint is ship stability, (Maki *et al.,* 2011) developed a genetic algorithm for weather routing that takes into account ship stability. According to (Jovanoski & Robinson, 2009) parametric rolling and slamming are dangers to ship stability. Parametric rolling can occur during certain wave conditions when the ship is forced into a lateral rolling motion. Maximum roll angles of up to 40° have been reported by (France *et al.,* 2003). This is especially critical with deck-stowed containers and it also lowers the speed limit of the ship which must be considered in weather routing.

2.2. Prediction uncertainty in weather routing

Variables that describe structural limitations of the ship (parametric rolling, speed limits, bending moment, etc.) are calculated based on the NWP. Therefore, uncertainty estimates are required for predictions of each of these variables (not just of the fuel consumption and travel time) to support logistical concepts such as *slow steaming, virtual arrival* (i.e., dynamic ship speed management) and *fleet planning*.

Weather routing with WaSP As mentioned before a WaSP is a propulsion system that combines a conventional combustion engine with an additional wind propulsion system (WPS). To increase the efficiency of a wind propulsion system (WPS) the shipping route may detour off the shortest route (great circle) to exploit favorable weather conditions. Even detouring far off the shortest route can lead to fuel savings if the WPS significantly reduces the ship motor energy. Detouring off the shortest route requires taking into account NWP uncertainty which is why NWP uncertainty influences the fuel saving potential of a WaSP system (eq. 2.1).

$$\text{NWP uncertainty} \leftrightarrow \text{detouring} \leftrightarrow \text{fuel saving} \qquad (2.1)$$

2.2.1. Wind propulsion systems (WPS)

In 2015 the International Maritime Organization introduced Emission Control Areas (ECA) in "regulation 14". Ships are only allowed to burn fuels which contain no more than 0.1 % sulfur in ECAs. In general the allowed sulfur content is going to decrease from 3.5 % to 0.5 % by the year 2020. Therefore, in the long run fossil fuels are expected to become more expensive than today's mostly used heavy fuel.

One of the measures, that shipping companies are investigating to limit the foreseeable cost explosion, are WPS. The basic idea is that a ship can lower its engine power by the forward propulsion of the WPS. Another option is to increase the ship speed with the WPS. Yet the second option also reduces the forward propulsion of the WPS since the apparent wind becomes more forwardly with increasing ship speed whereas the WPS is most effective with lateral or stern wind. The minimum difference between the direction of the apparent wind and the direction of the ship heading depends on the technology of the WPS.

This section describes models for three different WPS which have actually been constructed and are installed on different ships. The propulsion power generated by each WPS is calculated based on its dimensions, the true wind speed and the true wind angle. The apparent wind speed *AWS* (eq. 2.2) and the apparent wind angle *AWA* (eq. 2.3) are obtained by vector calculation with the ship speed and heading.

AWS = apparent wind speed (2.2)
(results from the true wind and the ship's head wind)

AWA = apparent wind angle (2.3)
(angle between direction from where the wind blows and the heading direction of the ship)

Since the wind data refer to a height of 10 m above sea level, each value is converted to the operational height h of the WPS by eq. 2.4. The equation is an alternative approach by Aschenbeck *et al.* (2009) to the logarithmic wind profile that is described in sec. 3.3.1.

$$AWS_h = AWS \cdot 0.145 \, \text{m} \cdot (4.6 + \ln(h)) \qquad (2.4)$$

The following sections describe the normalized propulsion force FN of different WPS. FN is calculated based on AWA. The term *normalized* means that FN defines the propulsion force of a specific wind angle independently from the wind speed or dimension of the WPS. The basic equation to model the resulting force of a certain wind propulsion system at a ship speed V_s has the general form of F_{WPS} in eq. 2.5 that is further described in Bentin *et al.* (2016).

$$F_{WPS} = V_s \cdot \frac{\rho}{2} \cdot AWS_h^2 \cdot A \cdot FN \qquad (2.5)$$

A is a scaling parameter for the size of the wind propulsion system, ρ is the air density. The power P_{WPS} obtained by the wind propulsion system reads as

$$P_{WPS} = F_{WPS} \cdot V_s \qquad (2.6)$$

and the corresponding power reduction of the ship engine is defined by eq. 2.7. η_T denotes the total efficiency to generate propulsion energy from motor energy. A concise introduction to this subject is given by (MAN, 2011).

$$\Delta P_{WPS} = \frac{P_{WPS}}{\eta_T} \qquad (2.7)$$

2.2.1.1. Sail

Different kinds of sails have been in use for thousands of years and in contrast to other WPS (described in secs. 2.2.1.2 and 2.2.1.3) the interaction of multiple sails with the wind is well known. With a suitable profile a sail can be very effective even when the angle between true wind and heading direction is narrow (which is often the case for modern freight ships with high vessel speeds). A new approach is the so called DynaRig, a concept of an adjustable sailing system invented by Prölls (1970) based on work by Wagner (1966).

A modern realization of the DynaRig is the Maltese Falcon in fig. 2.2, a 66 m long three-mast yacht which was completed in 2006. It possesses three robust rotational masts made of carbon fiber, each equipped with 5 automatically deployable sails. Based on sensors and a computer software the sail can be adjusted to individual wind conditions. Another example is the "Shin Aitoku Maru" which was built in 1980 with a wingsail. It was the first cost-effective motor ship with WaSP since the end of the traditional sail cargo shipping era which was shown by Ouchi *et al.* (2011).

Normalized propulsion force This section describes a model of a DynaRig installation with three masts ($N_M=3$) similar to the Maltese Falcon in fig. 2.2. Appendix A.3 contains a table with the parameter values. Interference between sails is not considered in the model, i.e., it is assumed that the masts are installed far enough away from each other. The normalized propulsion force of the DynaRig in eq. 2.8 depends on the propulsion coefficient Ca and the resistance coefficient Cw (both are dimensionless) which change with the wind conditions and the sail angle.

$$FN = Ca \cdot sin(AWA) - Cw \cdot cos(AWA) \qquad (2.8)$$

The polar chart in fig. 2.1 shows *FN* for different *AWA*. Obviously, the DynaRig is most effective for lateral and stern apparent wind angles.

Figure 2.1: *Normalized propulsion force of the DynaRig for different wind angles*

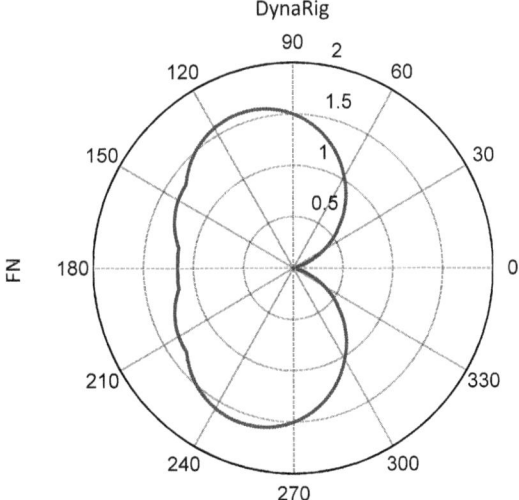

Figure 2.2: "The Maltese Falcon super yacht has been sailing with the DynaRig system for years. In its first 12 months of operation the ship sailed 23,310 nautical miles—including two Atlantic crossings—deployed 12,179 individual sail sets, achieved a top speed of 24.9 knots and used wind propulsion for 61 % of all its time spent at sea."[1].

---

1 Source: http://www.ship-technology.com/features/featurefossil-fuel-free-environment-shipping/featurefossil-fuel-free-environment-shipping-4.html (accessed 23.12.2015).

2.2.1.2. Kite

A kite-sail, or simply kite, consists of light robust tissue and pulls a ship by a rope from the kite to the foreship. To ensure stability the kite constantly flies an "eight"-like figure. It flies at heights of up to a hundred meters and it is (semi-)automatically operated by an electrical controller. The flying speed of the kite causes a higher apparent wind speed for the kite than experienced at the ship which increases the efficiency of the kite significantly. According to (Schlaak et al., 2009) a 600 m² kite-sail saves about 30 % propulsion energy on certain routes (e.g., North America to Europe). But there are still some open technology questions, e.g., safety considerations.

The model of the kite in fig. 2.3 describes the kite similar to the one which was installed on the BBC SkySails and is described in Aschenbeck et al. (2009). The model parameter values are listed in appendix A.4. The electric drive that operates the kite consumes energy which is subtracted from the propulsion energy that is saved by the kite.

Figure 2.3: *Normalized propulsion force of a kite-sail for different wind angles. The propulsion force is maximized for stern apparent wind angles.*

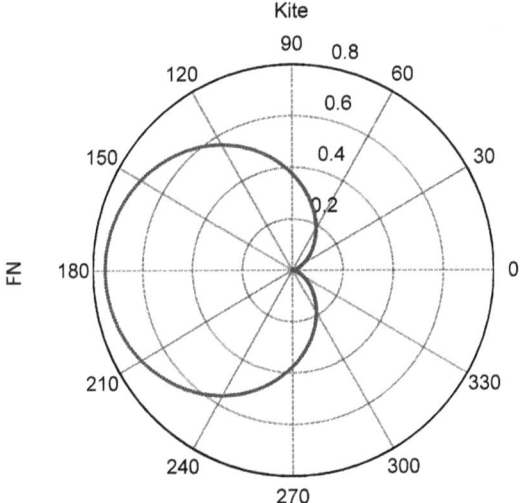

2.2.1.3. Flettner-rotor

The Flettner-rotor is a rotating cylinder which is installed on a ship. The E-Ship 1 in fig. 2.4 uses 4 Flettner-rotors. The incoming air flow is accelerated on one side of each Flettner-rotor and decelerated on the other side, similar as with a conventional sail. This causes a pressure gradient between both sides of the rotor according to Bernoullis equations. This so-called *Magnus effect* was first described by Magnus (1853).

In contrast to a sail the motor which rotates the Flettner-rotor consumes energy. The optimal rotation speed depends on the individual wind conditions and can be determined by an iterative algorithm. A press release by Enercon claimed that the rotors of the E-Ship 1 yield up to 15 % fuel savings compared to a conventional ship.

The E-Ship 1 was equipped with several special ship constructions and the savings refer to mean values on several different routes. The study was carried out in the North and Baltic Seas, the Mediterranean Sea and the North and South Atlantic Oceans. The model parameter values in appendix A.2 are similar to the dimensions of the Flettner-rotors installed on the E-Ship 1 in fig. 2.4.

Figure 2.4: *The picture shows the E-Ship 1 which was completed in 2010.[2] Four rotating Flettner-rotors aid the propulsion of the ship by exploiting mainly lateral wind forces.*

---

2 http://cdn2.shipspotting.com/photos/middle/0/8/1/1458180.jpg (accessed 09.08.2015).

2.2. Prediction uncertainty in weather routing

Normalized propulsion force The propulsion force of a Flettner-rotor for given wind conditions and rotation speed is calculated based on the work of (Wagner et al., 1985). The model of the Flettner-rotor differs from the models of the DynaRig and of the kite because the propulsion force FN in fig. 2.5 is not obtained with a specific sail angle but with an optimal rotation speed (depending on the wind conditions) of the Flettner-rotor(s) that is determined by an iterative algorithm.

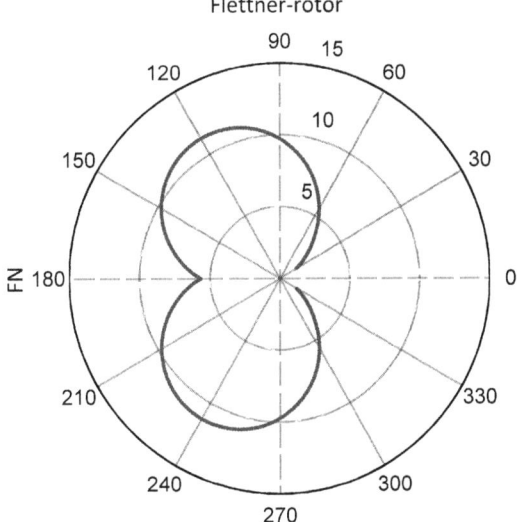

Figure 2.5: *The normalized propulsion force of the Flettner-rotor for different wind angles. In contrast to the DynaRig and the kite-sail the Flettner-rotor is most effective with lateral wind angles.*

2.2.2. Speed power curve

A WPS provides a portion of the propulsion power of a ship. The speed power curve P_{cw} in eq. 2.11 defines the required propulsion power to sustain a ship speed V_s in calm water, i.e., without taking into account additional wind and wave resistances. P_{cw} does also depend on the draft T of the ship. A linear model in eq. 2.10 is used to take into account T. The parameter values a_1 and a_2 in tbl. A.1 were fitted (based on on-board measurements) for a multi-purpose carrier (MPC), the "BBC Hudson".

The values of T and V_s in appendix A.1 represent mean values set for the simulation. It should be noted that eq. 2.11 is not valid under specific environmental conditions, for example in cases where high head sea generates a lot of frontal wave resistance on the ship, so that the ship engine does not generate enough power to sustain the required rate of rotation anymore. This model inaccuracy is not further considered here since the weather routing algorithm that is used here (described in sec. 6.2) is designed to exclude routes which travel through extreme weather situations.

$$b = \begin{cases} 3.5 & \text{if the ship is a MPC or Bulker} \\ b(T) & \text{if the ship is a Tanker.} \end{cases} \quad (2.9)$$

$$R_{cw} = a_1 \cdot T + a_2, \quad (2.10)$$

$$P_{cw} = R_{cw} \cdot V_s^b \quad (2.11)$$

2.2.2.1. Propulsion efficiency

To calculate the ship engine propulsion power that is needed to overcome wind and wave resistances (or that is saved by a WPS) the propulsion efficiency of the ship has to be taken into account. The required engine propulsion power depends on the properties of different ship components (tbl. 2.1) which transform the motor energy into propulsion energy.

Table 2.1: *Efficiency coefficients of the BBC Hudson as mean values for a travel speed of approximately 13 kn.*

name	symbol	value
open water propeller efficiency	η_O	0.62
hull efficiency	η_H	1.15
relative rotative efficiency	η_R	0.97
shaft efficiency	η_S	0.985
total efficiency (machine → water)	η_T	$\eta_O \cdot \eta_H \cdot \eta_R \cdot \eta_S$

2.2.3. Wind resistance

The wind resistance R_{Wi} is calculated with equations by Blendermann (1996) which he derived from experiments. The parameters for the wind resistance model are listed in appendix B. The wind resistance is calculated based on the angle ε between the ship course and the wind direction. ε is calculated by using vector calculation based on V_s, the ship course S_c, the direction of true wind W_d and the true wind speed V_w (eqs. 2.12–2.15).

$$\gamma = |S_c - W_d| \tag{2.12}$$

$$V_r = \sqrt{V_s^2 + V_w^2 + 2V_s V_w \cos\gamma} \tag{2.13}$$

$$\cos(\varepsilon) = \frac{V_s^2 + V_r^2 - V_w^2}{2 V_s V_r} \tag{2.14}$$

$$\varepsilon = \cos^{-1}(eq.\ 2.14) \tag{2.15}$$

The parameters that specify the ship silhouette exposed to the wind are listed in appendix B. The corresponding wind power P_{Wi} is obtained by multiplication of the wind resistance with the ship speed (eq. 2.16). Dividing P_{Wi} by the propulsion efficiency coefficient η_T (tbl. 2.1) yields the required ship propulsion power ΔP_{Wi} (eq. 2.17) to overcome P_{Wi}.

$$P_{Wi} = R_{Wi} \cdot V_s \tag{2.16}$$

$$\Delta P_{Wi} = \frac{P_{Wi}}{\eta_T} \tag{2.17}$$

2.2.4. Wave resistance

The wave resistance is calculated with software from the "Hamburgische Schiffs- und Versuchsanstalt" described by Blume (1977). The model parameters are listed in appendix B. Like the wind resistance model the wave resistance model is based on empirical data. The original `Fortran77` code has been converted to a Java implementation for this work within in the scope of the bachelor thesis of Traumüller (2014).

The ship model is parameterized with the dimensionless resistance coefficient CB which depends on the draft T. Analogously to the wind power the wave power is obtained by multiplication of the wave resistance with the ship speed V_s (eq. 2.18) and the required ship propulsion power to overcome the wave power P_{Wa} is calculated by dividing P_{Wa} by the propulsion efficiency coefficient η_T (eq. 2.19).

$$P_{Wa} = R_{Wa} \cdot V_s \tag{2.18}$$

$$\Delta P_{Wa} = \frac{P_{Wa}}{\eta_T} \tag{2.19}$$

2.2.5. Ship propulsion energy

The preceding four sections described models for the calm water (eq. 2.11), wind (eq. 2.16) and wave resistances (eq. 2.18) on a ship with a wind propulsion system (eq. 2.7). Together these models calculate the propulsion power of WaSP for specific wind and wave conditions, draft T and ship speed V_s. Adding together these values and multiplying with the travel time Δt yields the WaSP energy in eq. 2.20 where w_d^h is a NWP on a date d with a look-ahead horizon of h hours.

$$E(w_d^h) = (P_{cw} + \Delta P_{Wa} + \Delta P_{Wi} - \Delta P_{WPS}) \cdot \Delta t \qquad (2.20)$$

A ship route is defined by a departure date and a sequence of coordinates $\mathbf{R} = (pos_0, \ldots, pos_L)$ between the departure and the destination coordinate. The distance between two coordinates corresponds to Δt hours of travel time. The total predicted ship propulsion energy E_P (eq. 2.21) for a route with departure date d is obtained by summing the ship propulsion energy calculated with wind and wave data at each coordinate $pos_i \in R$.

Analogously, E_A (eq. 2.22) is the *true* ship propulsion energy for the route calculated with weather analyses instead of weather predictions. The iterative procedure to calculate E_P and E_A is shown in fig. 2.6. The prediction error of E_P is denoted $E_{P,Err}$ (eq. 2.23).

$$E_P(d, R) = \sum_{\substack{pos_i \in R \\ h = i \cdot \delta t}} E(w_d^h(pos_i)) \qquad (2.21)$$

$$E_A(d, R) = \sum_{\substack{pos_i \in R \\ h = i \cdot \delta t}} E(w_{d+h}^0(pos_i)) \qquad (2.22)$$

$$E_{P,Err}(d, R) = |E_P(d, R) - E_A(d, R)| \qquad (2.23)$$

2.2. Prediction uncertainty in weather routing

Figure 2.6: UML activity diagram showing the calculation of the ship energy consumption for a ship with an optional wind propulsion system (WPS). The ship energy consumption is calculated by splitting the route into L segments, yielding an energy function of travel time E_P (weather predictions) or E_A (weather analyses).

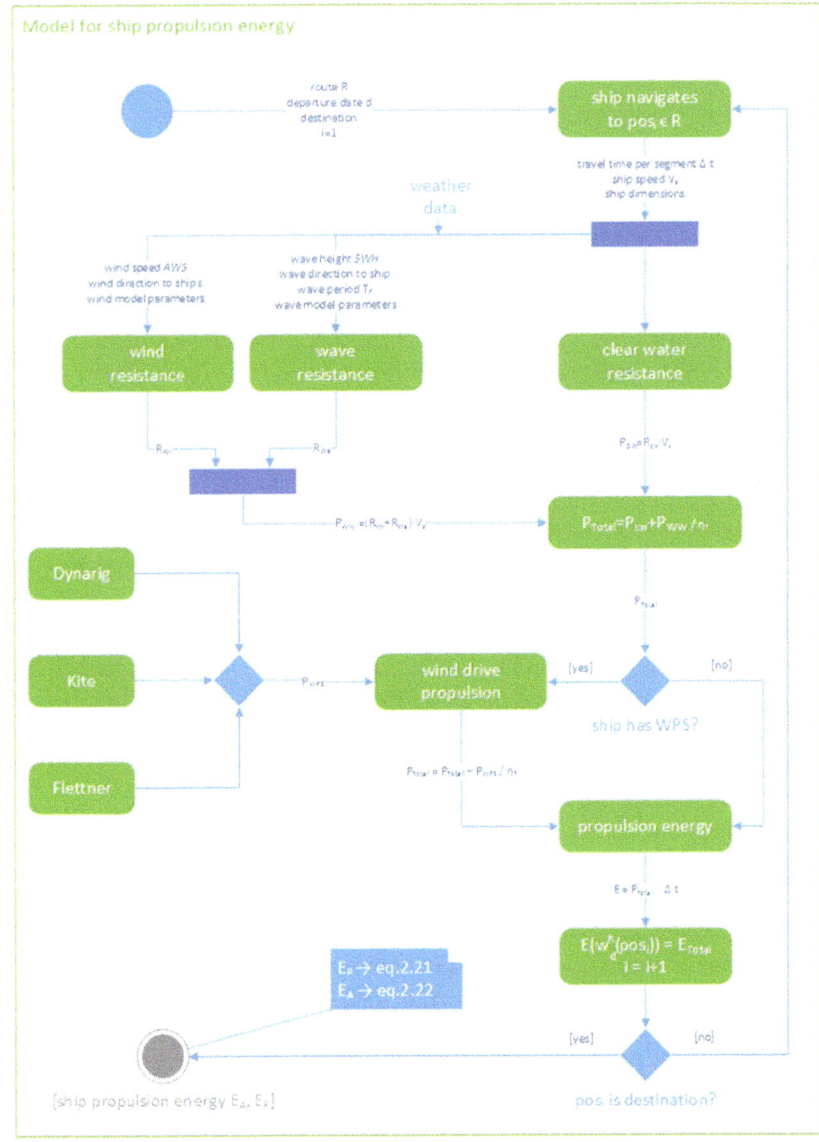

2.2.6. Prediction uncertainty with WPS

The impact of the WPS on ship propulsion energy prediction uncertainty depends on the type and dimension of the WPS technology. In general a WPS increases the effect of weather on the ship propulsion, hence making the ship propulsion less predictable. Therefore it is of utmost importance to reduce the prediction uncertainty.

2.3. Wind power generation versus weather routing

The assessment of the uncertainty of wind power predictions has been an active research area in the context of offshore wind farms. Several studies have investigated the link between prediction errors and characteristic weather situations. But the methods and results of these studies can only partially be transferred to the problem of prediction uncertainty for wind powered ships. Estimating the uncertainty of predictions for a (wind-powered) ship generally offers more challenges than estimating the uncertainty of a prediction for an offshore wind turbine because of several reasons:

Geographical non-stationarity If a weather prediction fails to predict the forward speed of a deep pressure area this does not necessarily affect the daily prediction error for a wind turbine. The wind turbine is stationary and will encounter the deep pressure area later. The ship on the other hand changes its position and can possibly avoid or encounter the deep pressure area dependent on its forward speed.

Prediction horizon For wind turbines the short-term predictions are of primary interest. The energy stock market relies on accurate power predictions 1–2 days ahead. Adjusting the power consumption of the regional industry to the available wind power production *(smart grids)* also requires reliable short-term predictions. But the short-term weather predictions (1–3 days) are often very reliable compared to the accuracy of weather predictions with a prediction horizon of up to seven days in ship weather routing.

Model complexity The power output of wind turbines depends non-linearly on the weather conditions. Nevertheless it has been shown that the relationship can be modeled analytically and in general stronger wind conditions correspond to higher power outputs since the wind turbines are adjustable to the wind direction. In the context of weather routing the relationship is ambivalent. Stronger winds can increase or decrease the ships wind resistance depending on the wind direction and the sphere of action of the wind drive (if a WPS is installed). In addition, stronger winds are usually linked to higher waves which can also influence the ship's propulsion power. The combined effect of strong wind and waves on a ship is a complex analytical problem.

Conclusion Several methods exist to estimate the prediction uncertainty in wind power generation (WPG). Yet the comparison above shows that there are fundamental differences to the problem of estimating the prediction uncertainty in weather routing. Therefore a new method has to be developed based on the approaches from the area of WPG.

2.4. Weather forecasting

So far it has been described how uncertainty in predictions in offshore applications origins from uncertainty in weather predictions. This section describes numerical weather forecasting and the reasons for prediction uncertainty. Weather forecasts are categorised by their forecast horizon (see tbl. 2.2).

Table 2.2: Common categories of weather forecast horizons

weather forecast	forecast horizon
very short-term	up to 12 hours
short-term	up to 3 days
medium-term	up to 14 days
long-term	up to one season

Since the mid-twentieth century the *numerical forecasting method* is the most accurate forecasting method and it is the only method which can be applied to any forecast horizon in tbl. 2.2 (except for long-term). It is based on complex physical models and substantial computational resources. Depending on the weather forecast horizon, different methods—described in (Ahrens & Samson, 2010, p. 408)— are available in order to supplement or validate the numerical weather forecast. A summary of these methods is given below.

- The *persistence forecasting method* projects the current weather without any changes to the future. The persistence weather forecast serves as a reference for the validation of other weather forecasts.
- The *trends forecasting method* computes the latest trends, e.g., a sequence of radar images showing precipitation data, and projects it onto the prediction horizon. The method is only applicable for very short-term weather forecasting.
- The *climatology forecasting method* uses long-term average weather parameter values as predictions. Although this enables arbitrarily long prediction horizons the predictions can be very inaccurate.

- The *analogue forecasting method* infers a weather prediction from historical weather situations which are similar to the current weather situation. Its performance depends on an accurate representation of actual weather conditions and on a rich data set with historical weather situations. Like the climatology method this method is based on statistical data.

The analogue forecasting method is not just useful to validate the numerical weather forecast but it can also be applied to compute prediction uncertainty of numerical weather predictions statistically, i.e., from the variance of historical prediction errors. To refine this idea the uncertainty in historical numerical weather predictions is assessed in chapter 3. The historical weather data are available in the form of a database with numerical weather predictions from the German Weather Service (DWD) which is described in the next section 2.4.1.

2.4.1. Numerical weather analyses and predictions

A numerical weather prediction (NWP) is run on a (super-)computer in two steps. The first step, called assimilation, computes the initial state of the atmosphere from all the available measurements and takes into account the historical data. Then, based on the equations for fluid dynamics and thermodynamics, the state of the atmosphere is projected for a limited time period into the future. Figure 2.7 shows that the average error of NWP for storm tracks of Atlantic storms has constantly decreased during the last decades. A rule of thumb states that NWP accurateness increases by one day per decade since the mid-twentieth century.

Figure 2.7: *Annual average official track errors of Atlantic basin tropical storms and hurricanes for the period 1970–2012, with least-squares trend lines superimposed by the American National Hurricane Center (NHC).*

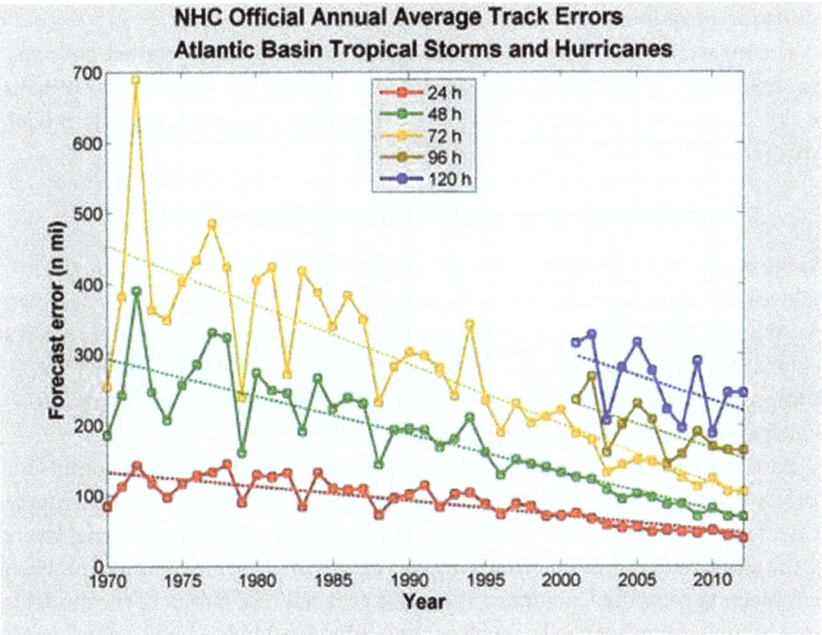

Source: http://www.nhc.noaa.gov/verification/verify5.shtml? (02.12.2015)

2.4.1.1. Weather analyses by the German Weather Service

The weather analyses are the result of the data assimilation process. This step is the model initialization. The data assimilation describes the state of the atmosphere based on observation data (or measurements) from a time period of 3 hours. Since the observations are irregularly spaced, a data assimilation procedure interpolates data for locations where no measurements are available. Each assimilation is calculated from the last assimilation that is modified based upon the latest measurements. The assimilated data are then used as a starting point for the numerical weather prediction. To predict the weather at a future point in time the latest weather analysis is used.

The 3-hourly weather analyses of the Global Sea Model (GSM) in the data set for the period from 2005 to February 2008 consist of very short-time forecasts (see tbl. 2.2) t+03h to t+12h (twice a day). Since February 2008 the GSM model includes a data assimilation and analyses at 00 UTC and 12 UTC with additional very short-time predictions t+03h to t+12h. The data assimilation is calculated purely with radar altimeter data, the measurements from buoys are only used for validation purposes.

The measurements from weather satellites, radiosondes in weather balloons, reconnaissance aircraft and pilot reports are used in the assimilation process of the "Extended Global Model" (GME). The GME is the atmospheric model, which forces the GSM model.

2.4.1.2. Weather predictions by the German Weather Service

As the prediction horizon increases, local weather predictions depend on the initial weather conditions that are farther away in time. According to the German Weather Service a 5-days numerical weather prediction (NWP) already requires a global forecast model like the "Extended Global Model" (GME) by the DWD. It imposes a global grid structure with an average grid spacing of 20 km (2014) which corresponds to a mean grid size of 346 km².

Based on the data assimilation, the GME forecast model produces outputs for surface pressure, horizontal wind components, temperature, specific contents of water vapor, cloud water and cloud ice, rain, snow and ozone on 60 model layers in the atmosphere from the surface up to a height of approximately 34 km. Each parameter is computed as a mean value of a grid cell. The output of the model is used as input for other models such as the Global Sea Model (GSM) with a mesh width of $0.75° \times 0.75°$.

The GME data in this work were mapped to the lower resolution of the GSM model. For applications in the North Sea and the Mediterranean Sea the DWD also maintains models with higher grid resolutions (Local Sea Model (LSM); $0.167° \times 0.1°$, Mediterranean Sea Model (MSM); $0.25° \times 0.25°$).

2.4.2. Limitations and trends in weather forecasting

There is a continuous competition between national weather services for the most accurate weather prediction. Improvements in hardware, physical models, simulation, sensor technology, science and "big data" led to a general one-day-per-decade improvement of European and US forecasting models (Greengard, 2014). This means that today's six-day predictions are generally of equal value as the five-day predictions 10 years ago (Ehrendorfer, 1997).

Much of the prediction quality depends on the available computing power. With more computing power the model resolution can be increased and a higher resolution means more accurate predictions that extend further out in time. But even with more powerful super computers the maximum possible weather prediction horizon with global models was proven to be limited to approximately two weeks (Lorenz, 1969). Currently, NWP horizons extend to approximately one week. A comparison between plots of weather predictions and weather analyses (i.e., the actual weather) is shown in fig. 2.8.

Local models Better prediction accuracy can be achieved by using local models for specific regions such as the LSM model by the German Weather Service. The grid resolution (area per cell) of the LSM is 33.6 times higher than the resolution of the GSM. It predicts sea conditions for the North and Baltic Seas—and for the Adriatic Sea, at the request of the German *Bundeswehr* which was engaged in the Balkan Wars at this time.

In general, local models with a higher resolution can describe structures that are too small for the coarser resolution of a global model. These structures are called eddies and can be calculated with *Large Eddy Simulation*. The size of eddies varies between 100 m and 1000 m (smaller structures are considered to be turbulences). However, a higher model resolution only improves the model accuracy as long as measurements and observations are available to fill the model cells with data.

Chaotic nature of weather Edward Lorenz discovered the chaotic nature of weather in 1961. In order to show the chaotic nature of the weather he simplified the known atmospheric model with 12 differential equations for air pressure, temperature, wind velocity, etc. The simplified model could only describe few and large-scale atmospheric patterns, but Lorenz used it to show that the state of weather changes between what he called "strange attractors".

More specifically, he discovered that the weather temporally remains in some characteristic state and then – sometimes suddenly and unpredictably – changes into another state. This system behavior is observable for example during changes between NAO+ and NAO- in the North Atlantic region.

The chaotic nature of weather is also referred to as the "butterfly effect", because for modern weather prediction models it implies that even small measurement errors of the initial state of the atmosphere can lead to huge prediction errors after some time. And there exist vast areas, e.g., the oceans, with only very few measurements available (except for satellite data).

2. Uncertainty in Wind Power Generation and Weather Routing

Figure 2.8: Comparison between predicted (l.) and true (r.) wind speeds. The ship route (black dots) has been optimized based on the predicted weather. The yellow dot marks the position of the ship at the given point in time.

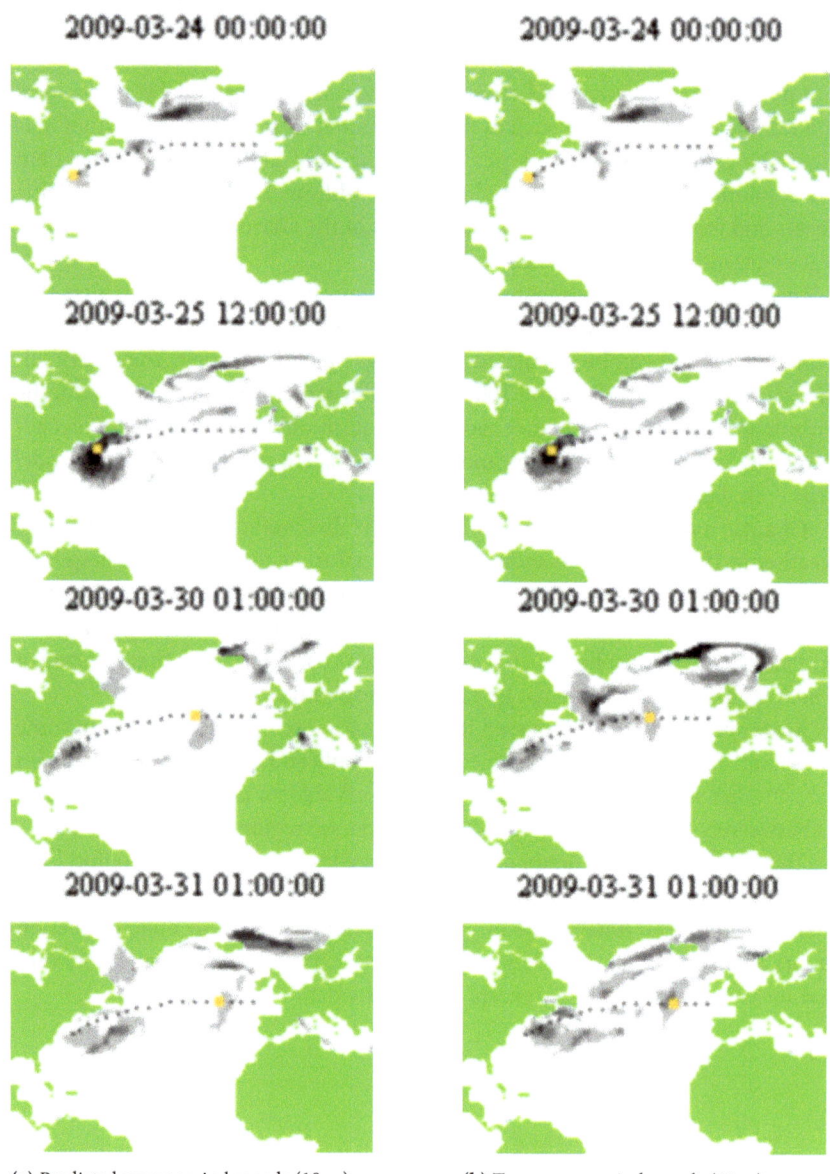

(a) Predicted average wind speeds (10 m) (b) True average wind speeds (10 m)

To estimate the uncertainty in predictions it is necessary to gather information from the data which indicates the level of chaos in the current weather situation. Chapter 3 describes a search for such data through the assessment of historical prediction uncertainty.

3. Statistical Patterns of Uncertainty in Weather Predictions

Abstract: *Chapter 2 described reasons and consequences of weather prediction uncertainty in two offshore applications. Therefore, this chapter assesses the accuracy of historical weather predictions in search for patterns that can be used to estimate prediction uncertainty. For this purpose, the historical weather prediction accuracy is calculated conditionally on individual months, prediction horizons and locations for the Northern and Baltic Seas and for the North Atlantic Ocean. Patterns are found with regard to long-term average prediction accuracy but these patterns vary significantly every year.*

Contents

3.1. Prediction error metrics 29
3.2. Predictions for the North Atlantic Ocean 30
    3.2.1. DWD wind and wave predictions..31
    3.2.2. Regional and seasonal prediction uncertainty 33
3.3. Predictions for the North and Baltic Seas 34
    3.3.1. FINO measurements...34
    3.3.2. Prediction uncertainty..37
    3.3.3. Wave prediction uncertainty 37
    3.3.4. Wind prediction uncertainty 42
    3.3.5. Conclusions..47

3.1. Prediction error metrics

Let $\hat{\mathbf{z}} \in \mathbb{R}^N$ and $\mathbf{z} \in \mathbb{R}^N$ be two vectors (predictions and observed values) where $\hat{z}_i$ denotes the i-th element of $\hat{\mathbf{z}}$ and z_i denotes the i-th element of $\mathbf{z}$ with $0 < i \leq N$. $\hat{z}_i$ is the predicted value for z_i. The signed difference Δz_i between $\hat{z}_i$ and z_i is defined by eq. 3.1.

$$\Delta z_i = \mathbf{z}_i - \hat{\mathbf{z}}_i \qquad (3.1)$$

The mean absolute error (MAE) between $\mathbf{z}$ and $\hat{\mathbf{z}}$ defines the error magnitude and is defined by eq. 3.2.

$$MAE(\mathbf{z}, \hat{\mathbf{z}}) = \frac{1}{N}\sum_{i=1}^{N} |\Delta z_i| \qquad (3.2)$$

The mean signed error (BIAS) between $\mathbf{z}$ and $\hat{\mathbf{z}}$ is defined by eq. 3.3. It defines the difference between the mean values of $\hat{\mathbf{z}}$ and $\mathbf{z}$.

$$BIAS(\mathbf{z}, \hat{\mathbf{z}}) = \frac{1}{N}\sum_{i=1}^{N} \Delta z_i \qquad (3.3)$$

The root-mean-square error (RMSE) between $\mathbf{z}$ and $\hat{\mathbf{z}}$ is defined by eq. 3.4. It emphasizes large prediction errors.

$$RMSE(\mathbf{z}, \hat{\mathbf{z}}) = \sqrt{\frac{1}{N}\sum_{i=1}^{N} (\Delta z_i)^2} \qquad (3.4)$$

If $\mathbf{z}$ and $\hat{\mathbf{z}}$ contain radial values eq. 3.1 is instead calculated by the smallest difference between both values in eq. 3.5.

$$\Delta z_i = ((z_i - \hat{z}_i) + 180) \bmod 360 - 180 \qquad (3.5)$$

This convention is used because the aim in this chapter is to assess the difference between expected and actual values for application purposes and not to validate the weather prediction model.

3.2. Predictions for the North Atlantic Ocean

There are only few direct offshore measurements available. Therefore, the weather predictions are validated by comparison with historical weather analyses. The prediction uncertainty is evaluated for different areas and seasons. For this purpose, the weather prediction uncertainty at eleven sites in the North Atlantic Ocean has been assessed (fig. 3.1). Each point (marked by a number) constitutes a historical local maximum of weather intensity to ensure comparability between the different sites, the lines are plotted for clarity. Predictions for the significant wave height (SWH) of the center of each region are evaluated for a time period from 2005 to 2010.

Figure 3.1: Partitioning of the North Atlantic Ocean into eleven regions

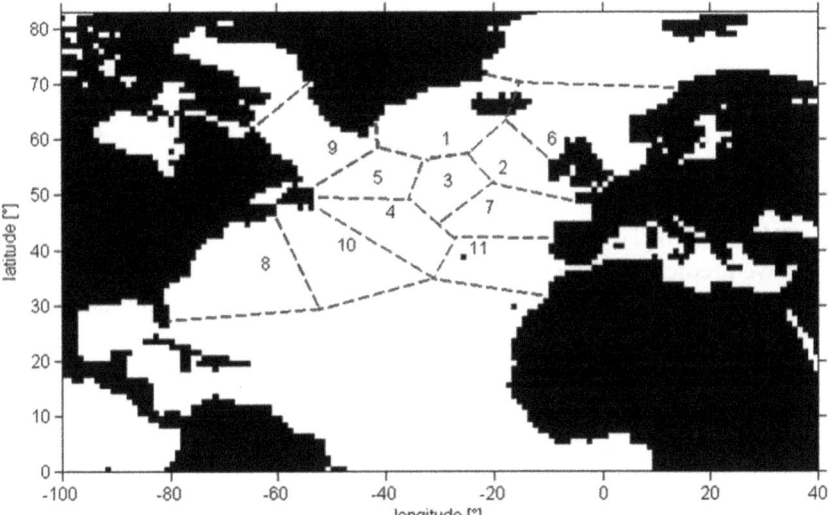

3.2.1. DWD wind and wave predictions

Each numerical weather prediction is based on the so called assimilation which defines the initial weather conditions based on all available weather observations. The wind and wave predictions by the German Weather Service (DWD) were made available for the time period from August 2005 to November 2010. It is noted that the predictions are mean values which describe whole regions to provide a coherent global representation. The measurements on the other hand are punctual. Therefore the prediction error is also linked to the grid resolution of the prediction model. Weather predictions in general and specifically wave predictions are used to select time windows for logistic and maintenance operations in offshore wind parks and oil platforms. Weather ship routing is another important application for wind and wave predictions. The predictions can be validated with local measurements from buoys, radars or laser measurement devices *(ground truth data)*. Another method is to validate the weather predictions with weather analyses if no measurements are available.

Wind predictions: Extended Global Model (GME) The wind speed (ms^{-1}) and wind direction (°) describe mean values 10 m above sea level. The wind predictions were generated by the Extended Global Model (GME). The GME is one out of 14 global weather prediction models. It operates on triangular grid elements with a resolution of 346 square kilometers. Four times a day all meteorological reports available within a few hours of the datum time (known as hor_0) are assimilated by the model's analysis scheme to define conditions at the datum time. This assimilation describes the current condition. The prediction is calculated based on this assimilation in 3-hourly intervals out to 174 hours, i.e., more than 7 days. The wind predictions are for example used by wind park operators to compute the wind park output for individual weather conditions. In 2015 the GME has been replaced by the ICON model.

Wave predictions: The Global Sea Model (GSM) The GSM provides predictions between [72°N, 72°S] with a resolution of 0.75° × 0.75° for multiple sea wave parameters (a complete list is given in appendix D):

- significant wave height (SWH)
- mean direction of wind waves (MDWW)
- mean direction of primary swell (MDPS)
- mean wave period (TP)

The significant wave height (SWH) is defined as the mean wave height in meters (trough to crest) of the highest third of all individual waves. The GSM is purely driven by computations from the global atmospheric model GME and a 12-hourly hind-cast from radar altimeter data. The hind-cast reconstructs past wave data for time periods when no observations were available.

Additional information about sea ice by the BSH (Federal Maritime and Hydrographic Agency) is used in the LSM to compute the wave energy in areas covered by sea ice. The wave analyses are calculated twice a day at 00:00 UTC and 12:00 UTC. The wave predictions are calculated based on the wave analyses for a look-ahead time of up to 174 hours, i.e., more than one week. This comes close to the typical travel time of a transatlantic container ship route. In 2012 the GSM model has been replaced by the GWAM model.

Table 3.1: DWD weather forecast models (http://www.dwd.de, August 3rd 2016)

model area		horizontal resolution	wind input data	spectral boundary data	forecast horizon
GWAM	Global	0.25°x0.25°	ICON	-	t+174h
EWAM	Europe	0.05°x0.10°	COSMO-EU	GWAM	t+78h
CWAM	German coasts	≈900m	COSMO-DE	EWAM	t+48h

3.2. Predictions for the North Atlantic Ocean

The DWD operates German, European and Global wind (COSMO-DE, COSMO-EU, ICON) and wave models (CWAM, EWAM, GWAM) which are shown in tbl. 3.1. The local models have higher resolutions than the global models. The GWAM has a maximum forecast horizon of 174 hours and provides boundary data for the European model EWAM which in turn provides boundary data for CWAM.

3.2.2. Regional and seasonal prediction uncertainty

Figure 3.2 shows the prediction RMSE (eq. 3.4) for the significant wave height in the eleven different areas (compare fig. 3.1) of the North Atlantic Ocean. Apparently, the areas 1, 2 and 6 in the vicinity of the trough of the Icelandic low pressure are showing the highest prediction error during winter. The prediction uncertainty varies geographically and seasonally. High storm risk in the Northwest Atlantic is contrasted with low storm risk in the Southeast Atlantic. The figure shows a distinct two-seasonal pattern with high prediction uncertainty occurring during the winter period between September and March.

Figure 3.2: Average monthly 120-hours prediction errors for the significant wave height (SWH) in eleven different areas in the North Atlantic Ocean.

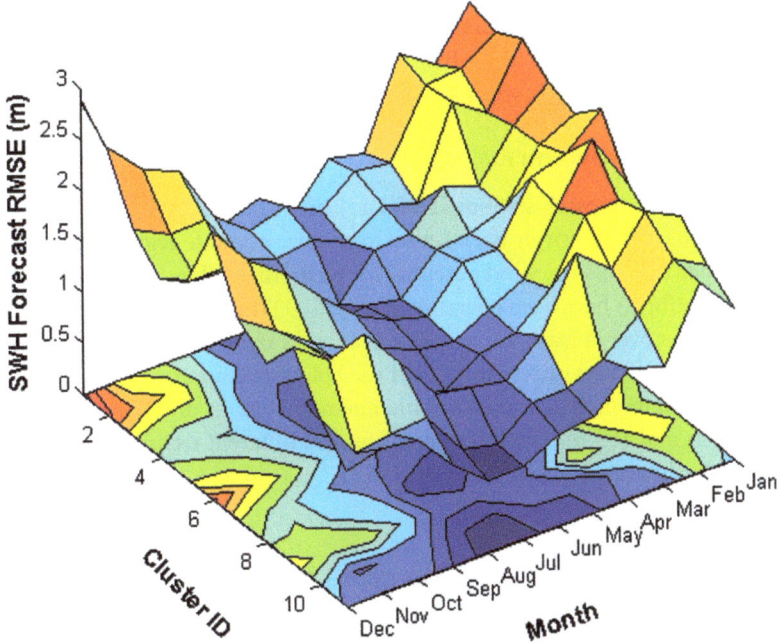

The next steps are to validate this result with measurements instead of weather analyses. To this end, the prediction uncertainty is assessed in the North and Baltic Seas where measurements are available. The measurements are supposed to be more accurate than weather analyses since the calculation of weather analyses is prone to the same causes as the calculation of weather predictions. Another objective is to assess the annual variation of the prediction uncertainty.

3.3. Predictions for the North and Baltic Seas

For the North Atlantic Ocean (and oceans in general) mostly observations from ships or satellites are available. Several permanently installed measurement platforms are located in the North and Baltic Seas because the water depths are usually less than 200 m. In the course of the German *Energiewende* three research platforms in the North and Baltic Seas (FINO) have been installed between 2003 and 2009[3]. This section describes the assessment of wind speed and wind direction prediction accuracy in the North and Baltic Seas by comparing measurements from offshore observation stations in the sea with prediction data from the German Weather Service (DWD). The 3-hourly predictions have been linearly interpolated in time. The seasonal and geographical variations are discussed by evaluation of the prediction error and the bias of the prediction error for look-ahead periods of up to 174 hours. The wind conditions in the North Sea are expected to differ from those in the Baltic Sea due to the long wave fetches and the stronger impact of Atlantic wind storms in the North Sea (Tambke *et al.*, 2006).

3.3.1. FINO measurements

Since 2003 three research stations (FINO 1, 2 and 3) (fig. 3.3) in the North and Baltic Seas have successively been set up to acquire accurate regional wind data and in order to assess the wind potential for offshore wind farms. Although not all of the platforms have been operational since 2003 (tbl. 3.2), the selected measurement data cover at least one complete year and can be used to assess seasonal variation in prediction uncertainty. The FINO platforms are equipped with wind vanes, cup anemometers and ultrasonic anemometers in different heights between 30 m and 106 m. The instruments feature different technologies and are installed on different sides of the mast[4].

---

3   The FINO project is financially supported by the German Ministry for Economy and Energy.
4   http://fino-offshore.de/en/ (accessed 20.12.2015).

3.3. Predictions for the North and Baltic Seas

Figure 3.3: The FINO research platforms in the North and Baltic Seas (picture taken from Google Earth, 14. April 2014).

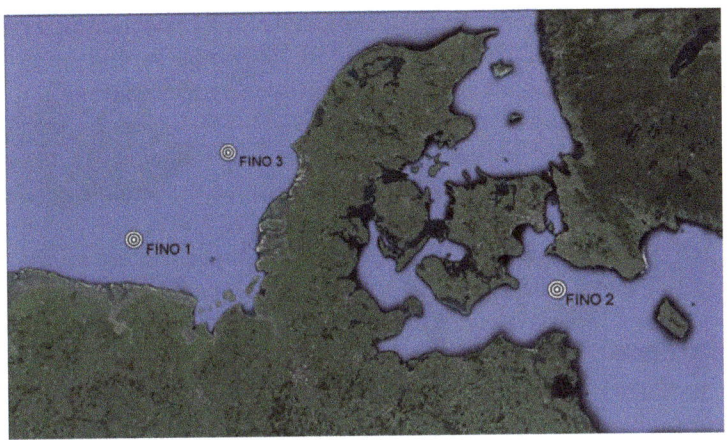

Table 3.2: Positions of FINO platforms

FINO ID	Longitude	Latitude	Operational since
1	6.58E	54.01N	2003
2	13.15E	55.00N	2007
3	7.15E	55.19N	2009

A study by (Beeken, 2010) found that the measurement mast causes average wind speed measurement errors of 1 %. In addition, the anemometer accuracy can slightly differ from the wind vane accuracy, depending on the weather conditions. Regardless, those effects should be insignificant compared with errors caused by the prediction model. Compared to the prediction date, the measurement data are set as "true/accurate", thus allowing to assess the accuracy of the DWD predictions.

To minimize the conversion error between measurements and predictions (that refer to a height of 10 m above sea level) the data from the instruments installed at the lowest height in tbl. 3.3 were chosen.

Table 3.3: FINO instruments and measurement heights

FINO ID	Wind speed	Wind direction
1	Wind cup (33 m)	Wind vane (33 m)
2	Wind cup (32 m)	Wind vane (31 m)
3	Wind cup (30 m)	Ultrasonic anemometer (60 m)

The logarithmic wind profile The measurements were collected in heights between 30 m and 60 m whereas the predictions refer to a height of 10 m. Therefore, the predictions are extrapolated to the measurement heights using a vertical wind profile. The approximation of the vertical wind profile depends on the temperature gradient (water-air) and the water surface's roughness length.

The logarithmic wind profile (Stull, 1988, pp. 378–379) is used to approximate the true wind profile in the atmospheric surface layer (up to 60–100 m). It is a simplification of the true wind shear but the measured data at different heights of the FINO stations can be described sufficiently well for the purposes of wind prediction validation. The logarithmic wind profile in eq. 3.6 defines the wind speed v_2 in height h_2 above sea level given a known wind speed v_1 at height h_1 at the same time. This accommodates for the fact that the wind speed generally increases with height.

The coefficient z_0 is the surface roughness. It depends on the wave forms and on the individual wind conditions. The theory about the relation between z_0 and the wind conditions is described by Charnock (1955). However, the theory does not fully apply in coastal waters since waves do not develop to their full height (depending on the depth of sea) and the wind direction changes during land-sea transition. For simplicity, a fixed value was used here. Regional reference values for z_0 can be found in the Wind Atlas Analysis and Application Program (WAsP—not to be confused with WaSP). Yet (Tambke et al., 2006) found that these values tend to be imprecise. For the Northern Sea they found $z_0 = 1\ cm$ to be adequate, which is also used here. Onshore deviations from the logarithmic wind profile are described by the Monin-Obukhov theory. But there is currently no generally accepted theory for offshore wind profiles. Reasons for this, apart from the ones mentioned above, are non-linear relationships between wind and waves, latent accumulation of water heat and the influence of onshore wind profiles in coastal areas (Lange et al., 2004).

$$v_2 = v_1 \cdot \frac{\ln \frac{h_2}{z_0}}{\ln \frac{h_1}{z_0}} \qquad (3.6)$$

Wind profile power law For the sake of completeness the wind profile power law by (Hellmann, 1914) is also mentioned here. It is based on empirical observations and takes into account external factors with the Hellmann exponent $k \in [0,1]$. (Kleemann & Meliß, 1993, pp. 248–249) propose values for k for variable values of surface roughness and uniformity levels of local wind profiles. If the atmosphere is turbulent, the wind velocities in the atmospheric surface layer are uniform, rough surfaces slow down the near-surface wind. According

3.3. Predictions for the North and Baltic Seas

to (Hau, 2008; Tambke et al., 2006, pp. 515–517), the wind power law and the logarithmic wind profile overestimate the wind shear with increasing height in the North Sea. In their study the logarithmic wind profile better approximates the true wind shear, therefore it is used here for the height conversion.

$$v_2 = v_1 \cdot \left(\frac{h_2}{h_1}\right)^\kappa \qquad (3.7)$$

3.3.2. Prediction uncertainty

The first part of the evaluation assesses the uncertainty of wave predictions at FINO 1, the second part investigates the wind prediction uncertainty at FINO 1–3. To compare the measurements with prediction values the 3-hourly wind predictions have been linearly interpolated before comparing them to the 10-minutely measurement intervals. The deviations between prediction values and measurements arise from inaccurate measurements, the weather model resolution in time and space, incorrect weather predictions and the approximation of the vertical wind profile.

3.3.3. Wave prediction uncertainty

The FINO 1 platform is equipped with multiple meteorological measurement devices. One of these devices is a state-of-the-art WaMoS II wave radar, which is installed on the downside of the FINO 1 platform. The wave conditions do not affect the performance of the wind turbines but can be critical during construction and maintenance operations. The comparison of WaMoS II measurements and measurements from a buoy by Borge et al. (1999) showed that WaMoS II measurements are highly accurate.

Also the WaMoS II measurements are supposed to be more suitable for the purpose of validating weather predictions, because both data describe mean values for a reference area in contrast to the local point measurements of a buoy. The accuracy of WaMoS II measurements is given as ±10 % relative to the true wave height and will be used as Ground Truth data (true value) here. The stationary WaMoS II measurements are used to validate the GSM wave predictions from 2008 by the German Weather Service (DWD)[5].

WaMoS II system The WaMoS II system is one out of several established wave radar systems which can be applied stationary as well as on a ship. On a ship

---

[5] The data were kindly been provided by Dr. T. Bruns (DWD).

the orientation of the installation is important but on the research platform the radar can be applied in each direction equally well. The system measures, digitalizes and saves a video stream of images from an X-Band radar in real-time. Normally, waves are causing undesirable noise (sea clutter) on radar images, which is removed by algorithms described by Borge *et al.* (1999). The WaMoS II system uses this sea clutter to compute directional wave spectra. These wave spectra are then processed to compute all relevant wave parameters such as the significant wave height, peak period and peak direction for both wind sea and swell.

In order to be operable the system needs to be calibrated once with measurement or wave analysis data. Figure 3.4 shows the data flow from the wave radar to one or multiple PCs in the WaMoS II system. For a wave radar measurement a sequence (time series) of 2-dimensional radar images is used as input. Subsequently, the sea wave clutter on the radar images is transformed into a spectral analysis via a 3-dimensional Fast-Fourier-Transformation (FFT).

Several algorithms compute the desired wave parameters from the spectral analysis. WaMoS II provides measurements as a 20 minute moving average. The data can be sent to a PC and forwarded via internet, LAN, NMEA etc. to other stations. The system only provides measurements for minimum significant wave heights of 0.5 m. Therefore, the statistical assessment of the wave prediction error could only be conducted for scenarios with sufficient wave heights.

Figure 3.4: Schema of the WaMoS II system and the data flow from the radar antenna to the user monitor(s) by (Jarabo-Amores et al., 2008, p. 37)

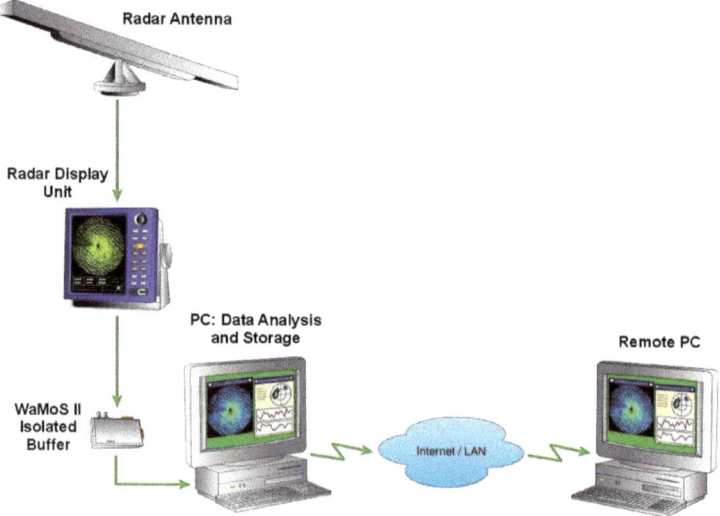

3.3. Predictions for the North and Baltic Seas

WaMoS II vs. GSM analyses Figure 3.5 shows errors for wave heights and directions (1 year) at FINO 1. Clearly, there are less values for wave direction prediction errors (MDPS, MDWW) since measurements of wave directions are sometimes not available. Differences between DWD wave analyses and WaMoS II wave measurements are interpreted as wave analysis errors. If the sea is calm it is not possible to detect wave directions for swell and wind sea. In 95 % off all cases the difference between significant wave height (SWH) analyses and measurements is 0.7 m or less. The mean wind wave direction analyses are less accurate than the mean swell direction (MDPS) analyses with an average error of approximately 50 °.

The accuracy of the wind direction analyses (MDWW) shows that an accurate calculation of the wave energy is not always possible if it is purely based on weather analyses. This can only be achieved by using accurate wind and wave measurements. But if the sea is calm, the WaMoS II often does not provide measurements for the swell direction (MDPS), either because the wave height is too low (<0.5 m) or because only one wave system is recognized. Therefore, it is sometimes not possible to accurately calculate the theoretical wave resistance on the ship because measurements are not available and weather analyses are not sufficiently reliable, especially for the wave direction.

Figure 3.5: Mean absolute error (MAE, eq. 3.2) of GSM wave data (see sec. 2.4.1.1), calculated by comparison with measurements from the WaMoS II wave radar (green) with FINO 1 data from 2008. Each point shows the difference between a 3-hrs wave analysis and 18 WaMoS II measurements.

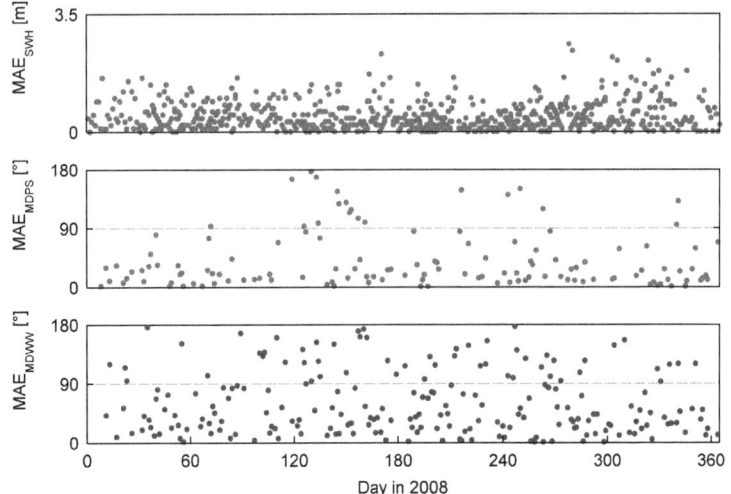

It should be noted, that in order to assess the impact of wave direction forecast errors for individual scenarios, the wave direction analysis has to be evaluated conditional on the wave height. However, fig. 3.5 still shows that the accuracy of the wave analyses does not significantly change between summer and winter.

WaMoS II measurements vs. GSM predictions Most applications depend on weather predictions rather than weather analyses. Therefore, daily predictions for 2008 have been compared with subsequent WaMoS II measurements. There are three possible reasons for deviations between DWD wave predictions and WaMoS II measurements:

- Weather analysis errors, e.g., because of sparse available offshore measurements.
- Algorithmic approximations of actual fluid dynamics in the weather prediction model, because the true dynamic equations are too complex for real-time calculation.
- Errors due to mean values for different areas: The wave radar computes a three-minutely moving average for an area with a radius of 4 km around FINO 1. The geographical resolution of the GSM is significantly lower: (0.75 · 34 km) · (0.75 · 40.5 km) = 775 km².

Figure 3.6: Absolute mean prediction errors between significant wave height (SWH), mean direction of wind waves (MDWW) and mean direction of primary swell (MDPS). The prediction error is the mean difference between DWD prediction values and WaMoS II measurements for a prediction look-ahead time of up to 174 hours. The data are daily predictions for 2008 at FINO 1 in the Northern Sea.

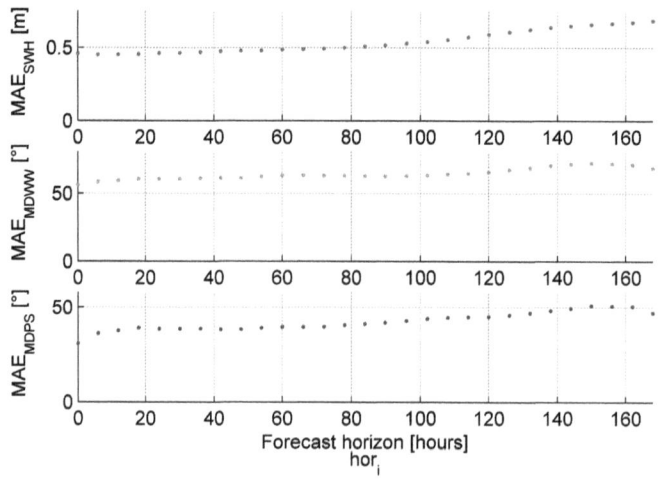

3.3. Predictions for the North and Baltic Seas

Again there is a difference between wave height and wave direction data. The average prediction error for the significant wave height in fig. 3.6 shows a linear increase for a look-ahead period of approximately 4 days (96 hours). For the remaining three days the average prediction error increases with a steeper slope. This change in prediction accuracy after 4 days is also visible for the prediction errors for the wave direction. This suggests a general decrease in wave prediction accuracy after approximately four days. The mean prediction error for the significant wave height (SWH) increases linearly from 0.5 m to 0.8 m, i.e., by a factor of 0.8 m/0.5 m=1.6.

The prediction error for the mean direction of wind waves (MDWW) increases by a factor of 65 °/50 °=1.3. This is less than the SWH factor, but only because the prediction accuracy is relatively low from the beginning. In addition, the prediction uncertainty fluctuates more than for SWH: The mean prediction error is almost constant for forecast horizons of up to 96 hours and then increases for longer forecast horizons of up to 168 hours.

Figure 3.7: *Mean prediction error for the wave height and wave directions for predictions in 2008 with a prediction horizon of hor_k (k = [0, 168]) hours. The mean absolute error (MAE) is the monthly average of absolute differences between predicted values and measurements.*

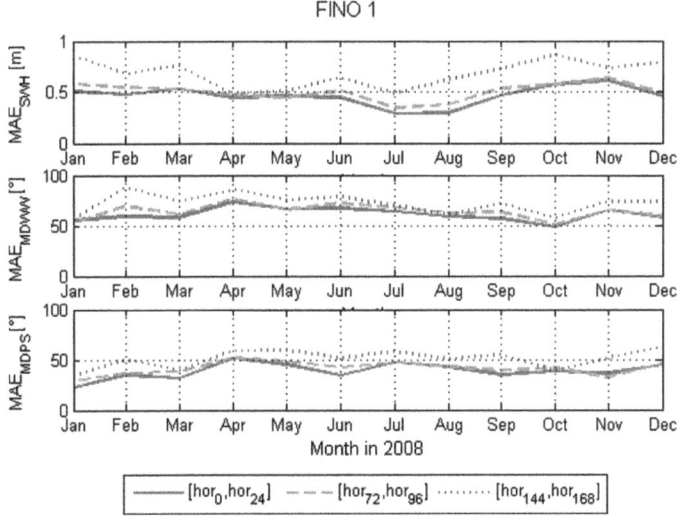

Forecast accuracy exhibits a seasonal pattern for SWH predictions which could not be observed for wave direction (MDWW, MDPS) predictions. The absolute deviations between predictions and measurements increase during winter. This can be explained with higher average wave heights during winter caused by North Atlantic deep pressure areas crossing the North Sea. Another observation is that errors in the analysis have a visible effect on the subsequent prediction. The evaluation of monthly prediction errors in fig. 3.7 shows that the seasonal pattern in the North Atlantic ocean (fig. 3.2) also exists in the German North sea, but only for the significant wave height. The pattern is slightly reversed for the prediction error of the mean direction of wind waves (MDWW). Prediction uncertainty for the mean direction of swell (MDPS) does not follow a seasonal pattern which is shown by the different average prediction errors in January and December for the data sample. It should be noted that the MDPS statistic is calculated with less data since measurements were often not available.

Again it is shown that prediction accuracy does not significantly degrade for a look-ahead period of up to 96 hours. The seasonal SWH pattern can be explained with more severe weather during winter which increases the potential for higher prediction errors. The most reliable wave predictions can be expected during July and August if the prediction horizon does not exceed 4 days.

3.3.4. Wind prediction uncertainty

Historical prediction accuracy for mean wind speed at 10 m above sea level (MU10) and mean wind direction at 10 m above sea level (MDWI) is assessed in this section. Aside from satellite data, only few direct measurements from the open sea are usually available for the initial weather assimilation process. Incorrect or missing initial measurement data is a main reason why the wind analysis can differ from the initial true wind conditions. Yet in areas where sufficient measurements are available, the wind speed analyses are often very accurate as depicted in fig. 3.8.

Figure 3.8: Comparison between wind speed measurements at FINO 1 and DWD wind speed analyses (hor$_0$) *in April 2010.*

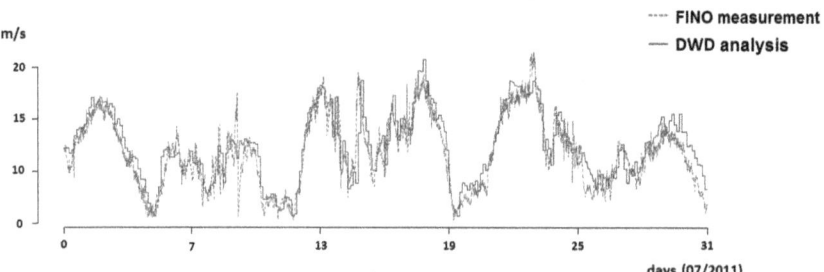

3.3. Predictions for the North and Baltic Seas

The large error ($\approx 7 \frac{m}{s}$) at day 9 is obviously a data error in the measurement data. This shows that measurement data need to be error-checked and smoothed, for example by using a Kalman filter. The bias at the end of the time period could be related to a wind direction where the measurement mast distorted the measurements.

Wind prediction bias As shown in fig. 3.8 a model bias can increase the average prediction error. Figure 3.9 shows that there is an average wind speed prediction bias of up to 1 ms^{-1} at FINO 1. The average bias of predictions for the wind direction with a look-ahead time of up to 24 hrs can even reach 40 ° in November. Here the error actually decreases with hor_i which indicates that the error is due to an inaccurate weather assimilation.

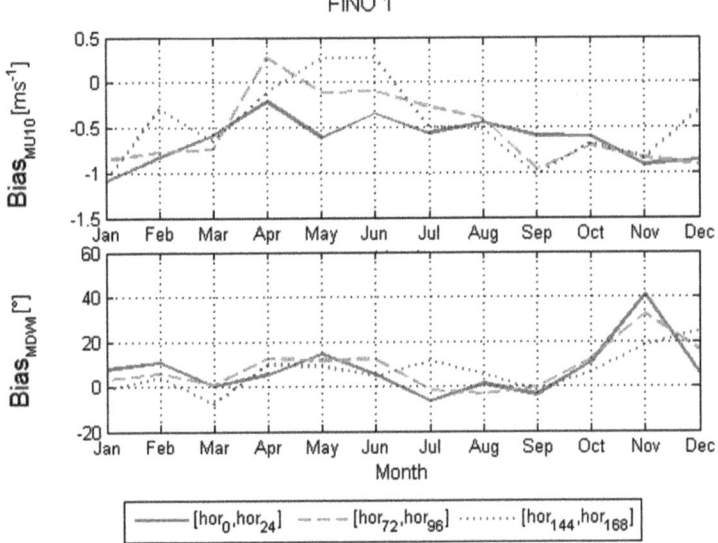

Figure 3.9: *GME wind speed bias at FINO 1. Daily wind predictions (2005–2010) were compared with subsequent measurements for a prediction horizon of one ([hor_0, hor_{24}]), four ([hor_{72}, hor_{96}]) and seven ([hor_{144}, hor_{168}]) days.*

Average wind prediction error A similar seasonal pattern as in fig. 3.2 (North Atlantic) is shown in fig. 3.10 (FINO). The mean-absolute prediction errors (MAE) for the wind speed increase during winter. The MAE of wind direction predictions spreads wider but shows no seasonal pattern. The pattern is less pronounced than in fig. 3.2 which can be explained with generally more severe weather on the ocean.

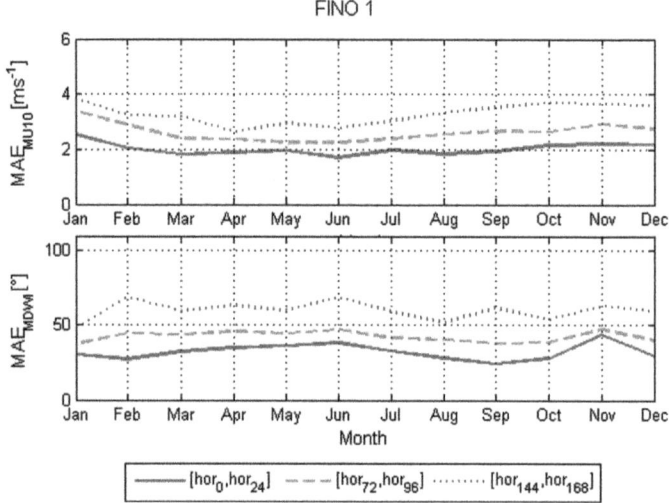

Figure 3.10: GME wind speed MAE for FINO 1. A comparison between the datum time prediction and one-week predictions. The t+k plot line shows the mean prediction error for prediction horizons of k hours.

Wind prediction error variance The wind speed prediction error distribution in fig. 3.11 complies with general knowledge that the wind speed assimilation error frequencies follow a Normal Distribution. The wind direction prediction error distribution in fig. 3.11 is more narrow and could be modeled with a student t-distribution. Its first and second moments, the expected value and variance, are the common parameters to describe the historical wind prediction uncertainty. Both distributions are almost symmetrical. The bias could be caused either by the location of FINO 1 in the GME grid resolution, by the incorrect measurement device calibration or by the influence of the measurement mast. If, for example, the variance of the distributions is used to estimate future prediction errors, the bias should be taken into account, too.

3.3. Predictions for the North and Baltic Seas

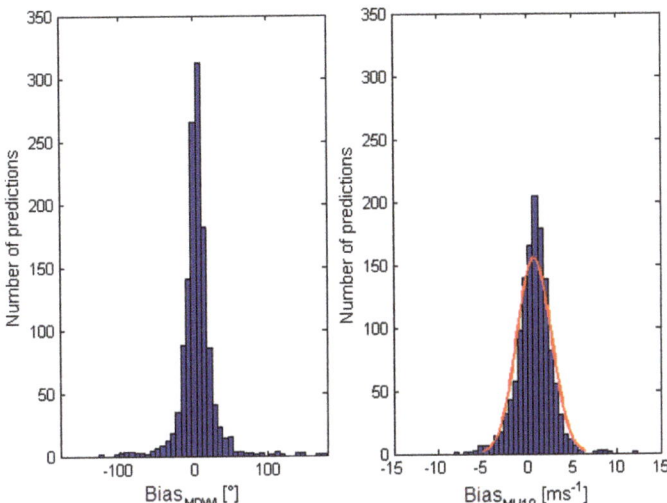

Figure 3.11: The histogram (blue) and a normal density function fit (red) show the deviation between DWD wind speed data (see fig. 2.4.1.1) and FINO 1 measurements (wind vane at 33 m) from 2005 to 2010.

Differences between North and Baltic Seas The distance between FINO 1 and 3 (both in the North Sea) is only 76 sea miles. So it does not surprise that prediction uncertainty decreases linearly almost equally for both sites as shown in fig. 3.12. Interestingly, the average prediction uncertainty at FINO 2 is also almost identical. The average wind speed prediction error (calculated based on fig. 3.12) is $\overline{MAE_{ws}} \approx 0.2 \cdot \frac{hor_i}{24} \frac{m}{s} + 1.7 \frac{m}{s}$ and the wind direction prediction error is $\overline{MAE_{wd}} \approx 6 \cdot \frac{hor_i}{24}° + 20°$.

The linear dependency of prediction uncertainty on hor_i indicates that the average prediction uncertainty can be inferred from hor_i. The leap of the prediction error shows the difference in accuracy between wind analyses and short-term wind predictions. In comparison with the wave prediction accuracy in fig. 3.6 the accuracy of wind predictions does not rapidly decrease after 96 hours. Here the increase of the average prediction error remains linear for the complete prediction horizon (except for the initial gap between wind analyses and wind predictions).

In summary, different statistical properties of prediction accuracy were observed which may be taken into account to estimate future prediction uncertainty:

a) a linear increase of the average prediction error proportional to the prediction horizon and
b) a seasonal difference between prediction accuracy for the wind speed and for the significant wave height and
c) a normal (or at least symmetrical and mono-modal) distribution of wind prediction errors.

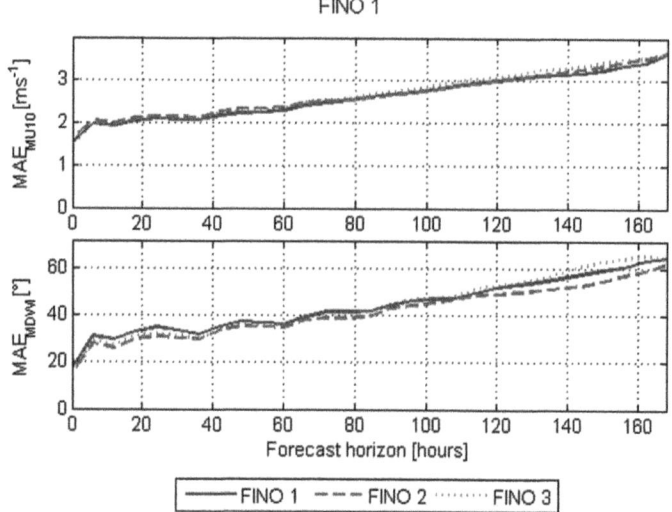

Figure 3.12: *Average prediction errors between 2005 and 2010 for wind speed at FINO 1-3*

Annual variation of prediction uncertainty The average prediction uncertainty increases linearly with the prediction horizon hor_i. Yet it was already shown that prediction accuracy varies between different months. Figure 3.13 shows that the linear dependency on hor_i also depends on the time of the year. During winter the average absolute wind speed prediction error MAE_{MU10} increases fast with hor_i. This statistic was obtained with data from a single year (2008) and it also shows that the pattern is less linear in this year compared with the statistic over 5 years in fig. 3.12. By adding the bias from fig. 3.9 to the spread in fig. 3.13 it shows that the average monthly prediction error may significantly deviate from the annual average prediction error. Furthermore, this not only accounts for the wind speed prediction error but also for the wind direction, wave height and wave direction errors.

3.3. Predictions for the North and Baltic Seas

Figure 3.13: Average prediction errors at FINO 1 for individual months calculated with historical data from 2008.

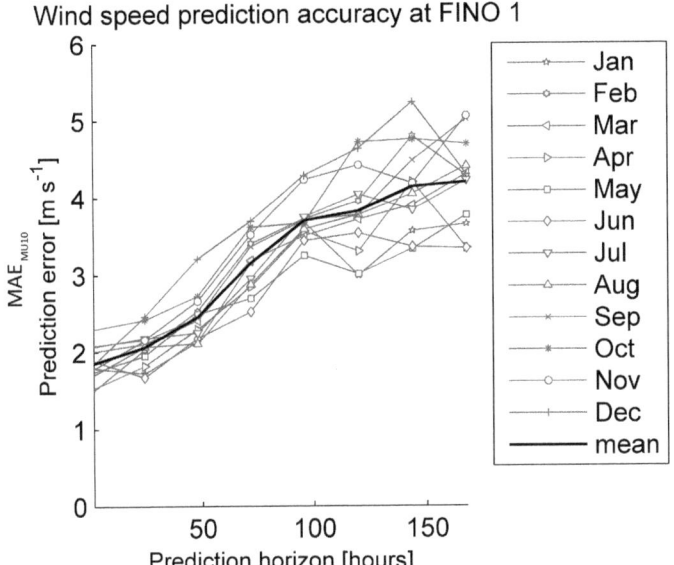

3.3.5. Conclusions

Statistical patterns No significant difference between prediction uncertainty in the North and Baltic Seas was found. It was shown that mean weather prediction errors for wind and wave predictions can be modeled as linear curves over the prediction horizon. In addition, a seasonal difference was observed of the mean prediction uncertainty of wind speed and significant wave height between summer and winter. The average uncertainty of predictions for wind and wave directions remains approximately constant throughout the year but is generally higher than the wind speed prediction uncertainty.

Uncertainty estimation To describe prediction uncertainty based on scenarios, the monthly wind speed prediction error plot provides useful information because the average prediction error depends non-linearly on the individual month m. The average prediction uncertainty for the wind speed and the significant wave height depends on m and the prediction horizon hor_i. In addition, the predicted energy E_P is also a predictor for the prediction uncertainty, since it is correlated with the actual energy E_A. There are different ways to calculate E_P, depending on the application. Section 2.2.5 describes how to calculate E_P for the propulsion energy of a ship. The calculation of E_P for a wind power plant is not further discussed here, since this is beyond the scope of this work. Depending on the technology and the wind conditions, the wings of a wind power plant start swinging as the wind speed increases and the plant needs to be turned off.

Table 3.4: *Predictors for the actual energy* E_A *(see sec.2.2.5).*

predictor	variable
predicted energy	E_P
month	m
prediction horizon (of E_P)	hor_i

With regard to the complexity of the problem of weather routing (see sec. 2.3) the information provided in tbl. 3.4 might still be too general in order to allow accurate estimations of the prediction uncertainty. Therefore, chapter 5 introduces a set of additional variables which are—just like E_P—calculated with the numerical weather prediction.

4. Estimation of Prediction Uncertainty

Abstract: *This chapter describes the state of the art of uncertainty estimation for weather predictions and weather-based energy predictions. First ensemble prediction systems (EPS) are discussed which estimate uncertainty in weather predictions by the spread of a set of possible weather scenarios. Then non-parametric statistical methods are discussed as alternatives to EPS. Quantile regression is selected as method for uncertainty estimation in this work since it is robust, fast and an established method in the field of uncertainty estimation of wind energy predictions.*

Contents

4.1. Theoretical and empirical models 49
4.2. Ensemble prediction systems 50
    4.2.1. Multi-model and multi-analyses ensembles 50
    4.2.2. Super ensembles 51
4.3. Statistical methods 51
    4.3.1. Probability density estimation methods 52
    4.3.2. Clustering 53
    4.3.3. Prediction intervals 54
    4.3.4. Quantile regression 55

4.1. Theoretical and empirical models

Theoretical, approximate and empirical models are available to estimate the uncertainty of predictions. If a proper theoretical model is available, it generally outperforms empirical models (Chatfield, 1993). Furthermore, in contrast to empirical models, the properties of theoretical models are verifiable by mathematical proof. Yet if a theoretical model misses important aspects of reality (e.g., rare extreme weather events such as El Niño or La Niña) or environmental changes, uncertainty estimates tend to be too narrow (Chatfield, 2001, pp. 475–494).

Empirical models can describe complex dependencies between variables if these are implicitly contained in the available data that are used to build the model. EPS (sec. 4.2) are a mixture between approximate and theoretical models whereas the non-parametric regression methods in sec. 4.3 belong to the class of empirical models.

4.2. Ensemble prediction systems

It was shown by Hoffschildt et al. (1999) and others that ship routing can benefit from ensemble forecasts. Some weather services provide ensemble prediction systems (EPS) to show the uncertainty of weather predictions. Most of the current systems (e.g., Zhi et al. (2012)) consist of a single model and a set of multiple perturbed initial weather conditions. Thus the EPS assume a perfect model and errors in the initial data assimilation (sec. 2.4.1).

Another approach is to estimate model uncertainty by *stochastic physics* or variable physical parameterization. The general aim is to estimate the uncertainty of the two *typical* error sources in weather predictions: An incorrect assimilation of the initial weather observations and errors in the prediction model, for example through the approximate mathematical methods to solve the differential equations in the model. Due to the chaotic nature of weather, which was described by Edward Lorenz (sec. 2.4.2), errors in the assimilation can have drastic consequences for the subsequent weather forecast. The spread of the predictions in the ensemble is an indicator for the uncertainty of the weather development. Ideally, the true weather then falls into the range of the ensemble prediction.

Often the spread of the ensemble is found to be too small, especially for prediction horizons of more than 10 days (Palmer et al., 2005). Multi-model ensembles are a common technique to address this issue.

4.2.1. Multi-model and multi-analyses ensembles

Multi-model ensembles (see for example Hagedorn et al. (2005)) calculate uncertainty based on the spread of predictions from different numerical weather prediction models. This is also sometimes called a "poor man's ensemble" which is rather misleading since the method requires predictions from multiple models. Since these are calculated with different models and different weather analyses the poor man's ensemble accounts for model and analysis uncertainty.

Stensrud et al. (1999) showed that both uncertainties need to be taken into account to increase the ensemble spread adequately. This is especially valid for short-term and medium-range predictions where perturbations in the analysis have most severe consequences. Medium range (up to five days) severe weather forecasting has always been a center of meteorological research, because human lives and economical values may depend on it.

A multi-analysis approach is the so called "lagged ensemble" which is calculated from predictions of a single weather service. The lagged ensemble consists of multiple succeeding weather predictions which predict the same future weather event. Significant differences between the predictions in the lagged ensemble indicate high analysis uncertainty. However, lagged ensembles do not account for model uncertainty.

4.2.2. Super ensembles

Super ensembles are also based on predictions from multiple models and require historical weather predictions and analyses for model calibration. Based on historical prediction uncertainty for each grid point and each variable, weights are assigned to the predictions of each weather model. The weights are optimized, e.g., with "multiple linear regression based super ensemble" (LRSUP) or "neural network based super ensemble" (NNSUP) techniques. Thus super ensembles give local weights to different prediction models based on the historical model accuracy in different geographical regions.

Super ensembles have been shown to significantly improve predictions compared with predictions from single weather models or average multi-model predictions, e.g., super ensembles yielded a 80 % lower precipitation RMSE (Mahmud, 2004). Regardless, despite their name, super ensembles do not indicate the spread or uncertainty of a prediction.

Current research The statistical interpretation of ensemble predictions is a subject of current research. Further research is necessary because an ensemble prediction is not equivalent to a probability distribution or prediction interval but only a set of possible scenarios (which do not necessarily cover the full spread of possible scenarios). The assessment of the quality of ensembles is a prerequisite to further improvements.

4.3. Statistical methods

In meteorological science ensemble prediction systems (EPSs) represent the state of the art to estimate uncertainty in weather predictions. Yet ensemble predictions require substantial computational resources which are sometimes not available. If ensemble predictions are not available, the prediction uncertainty may be deduced from historical weather prediction accuracy with a statistical approach – similar to the "analogue forecasting method" in sec. 2.4. Selecting an appropriate statistical method requires determining the requirements of the uncertainty estimates. There is one basic alternative in the literature (Ehrendorfer, 1997; Bremnes, 2004; Nielsen *et al.*, 2006; Pinson *et al.*, 2007a,b; Brabanter *et al.*, 2011; Bouckaert *et al.*, 2011; Bessa *et al.*, 2012; Zarnani & Musilek, 2013; Zarnani *et al.*, 2014) between approaches to estimate prediction uncertainty statistically:

- The first approach is to estimate a complete probability density function of the expected prediction value, e.g., with Parzen-Window Density Estimation. In a pre-processing step the data can be clustered to obtain a more specific density estimate.
- The second approach is to estimate prediction intervals which are accessible for example with non-parametric regression methods. Prediction intervals specify boundaries for the prediction error with a level of confidence α. Both approaches yield potential advantages and disadvantages which are described in the following sections 4.3.1–4.3.3.

4.3.1. Probability density estimation methods

From a statistical viewpoint the unknown energy E_A (eq. 2.22) is a random variable that is defined by its probability density function $p(E_A)$. The most important "state-of-the-art" methods to estimate $p(E_A)$ from a finite sample of observations are described in (Wang et al., 2011). If $p(E_A)$ is estimated conditionally on the value of E_p then the spread of $p(E_A)$ represents the prediction uncertainty of E_P. There are three different approaches available which either assume a distributional type of $p(E_A)$ (parametric methods) or infer the distributional type of $p(E_A)$ from the sample of observations (non-parametric methods).

Parametric methods are the best option if the distributional form of $p(E_A)$ is known, e.g., for modeling noisy normally distributed measurement. In complex real-world applications the density model is often unknown and therefore an a-priori model choice of the distribution type is not possible.

Non-parametric methods infer the distributional form from the data sample, i.e., the number of model parameters scales with the given data. A well-known non-parametric estimator is Parzen window density estimation. A well-defined kernel function (often a Gaussian kernel) is placed upon each data point. A common kernel center c and a common kernel width σ (also called the smoothing parameter) is chosen based on the data. For Gaussian kernels (eq. 4.1), σ is the standard deviation. The common kernel width is optimized based on "Silverman's plug-in principle" by (Silverman, 1986, p. 48) or the "leave-one-out estimator" by (Härdle, 2004, pp. 241–243).

$$N(E_A \mid c, \sigma) = \frac{1}{\sqrt{2\pi\sigma}} exp\left(-\frac{(E_A - c)^2}{2\sigma^2}\right) \qquad (4.1)$$

4.3. Statistical methods

To estimate the probability density function of wind power prediction uncertainty the Nadaraya-Watson kernel density estimator has been proposed by Juban et al. (2007) instead of the Gaussian kernel in eq. 4.1.

Semi-parametric methods, e.g., "finite Gaussian mixture models", do not assume a specific distributional form, i.e., they can approximate any continuous probability density function $p(E_A)$. In contrast to Parzen window density estimation the number of model parameters is fixed in advance. The probability density function $p(E_A)$ is approximated by a linear combination of mixture components (eq. 4.2), e.g., Gaussian kernels N. The "hat" in $\hat{p}(E_A)$ denotes that the function estimates the true function $p(E_A)$.

Bessa et al. (2012) showed that Gaussian kernels are not appropriate for wind power forecasting. They propose "beta kernels" for wind power prediction, "gamma kernels" for wind speed prediction and "von Mises distributions" for circular wind direction predictions instead.

Pinson (2012) proposes mixtures of generalized logit–normal distributions to estimate wind prediction error distributions. The kernel parameters c_m (kernel mean) and σ_m (kernel covariance matrix) can be estimated with the expectation-maximization (EM) algorithm by Dempster, A. P. et al. (1977).

N = sample size

M = number of mixture components

$$\hat{p}(E_A) = \sum_{m=1}^{M} N(E_A \mid c_m, \sigma_m), \text{ with } M \ll N \qquad (4.2)$$

4.3.2. Clustering

Density estimation is often combined with clustering. In a pre-processing step the data sample is clustered based on relevant attributes, e.g., the time of the year or the prediction horizon. Then a density estimate is computed for each cluster. Thus a conditional probability density estimate is obtained. For example, Fuzzy clustering by Zarnani et al. (2014) and hierarchical clustering by Pinson et al. (2007a) have been used to calculate conditional density estimates for wind power prediction uncertainty.

A disadvantage of the clustering method is that it involves further parameters (thresholds, distance metrics, cluster number etc.) which reduces the robustness of the uncertainty estimates, especially if the clusters do not contain enough historical data to estimate the underlying probability density function $p(E_A)$.

Current research Probability density estimation is an ongoing area of research. If methods are classified as non-, or semi-parametrical methods, the kernel parameter values such as the kernel width or kernel center, are selected based on historical data. The expectation-maximization (EM) algorithm and Silverman's plug-in principle are common methods to find suitable kernel parameters.

But the performance of these algorithms depends on the quantity and expressiveness of the data. If the application does not require a full probability density function it might be sufficient to estimate prediction intervals instead, since this requires less parameter tuning.

4.3.3. Prediction intervals

Prediction intervals show boundaries for the value of the dependent variable E_A with a certain level of confidence $\alpha \in [0,1]$. They do not define the probabilities for individual values of E_A. One method to calculate prediction intervals is by integration of the (conditional) probability density function. Therefore, prediction intervals describe the uncertainty in the prediction but they provide less information than a complete probability density function. A general discussion about prediction intervals is given by Christoffersen (1998).

In this work the point prediction for the wind energy is denoted by E_P and the true wind energy value calculated with true weather data (analyses data) is denoted by E_A. The meaning of a prediction interval $I^\alpha = [E_{P,Min}, E_{P,Max}]$ is that the true wind energy E_A is predicted to be within the limits of $E_{P,Min}$ and $E_{P,Max}$ with a probability of $100(1 - \alpha)$ % with $\alpha \in [0,1]$. If this is not the case then the prediction error I^α_E in eq. 4.3 is the distance between E_A and the limits of I^α.

$$I^\alpha_{Err} = \begin{cases} E_{P,Min} - E_A & \text{if } E_A < E_{P,Min} \\ E_A - E_{P,Max} & \text{if } E_A > E_{P,Max} \\ 0 & \text{else.} \end{cases} \quad (4.3)$$

In contrast to point predictions, the quality of prediction intervals depends on multiple criteria. The estimated prediction uncertainty I^α_U (eq. 4.4) (also called *sharpness*) defines the width of I^α.

$$I^\alpha_U = |E_{P,Max} - E_{P,Min}| \quad (4.4)$$

Reliability on the other hand is the percentage of E_A values that are covered by the prediction intervals. Let $I_{Test} = \{(I^\alpha, E_A)_i\}_{i=0}^N$ be a data set containing prediction intervals and associated true wind energy values. The reliability I^α_R is then defined by eq. 4.5.

$$I^\alpha_R = \frac{1}{|I_{Test}|} \cdot \sum_{(I^\alpha, E_A) \in I_{Test}} \begin{cases} 1 & \text{if } I^\alpha_{Err}(I^\alpha, E_A) \neq 0 \\ 0 & \text{else.} \end{cases} \quad (4.5)$$

4.3. Statistical methods

To benchmark prediction intervals with respect to sharpness and reliability with a single number, a so called skill score I_S^α was chosen. Gneiting & Raftery (2007) proved that the negatively oriented skill score in eq. 4.6 is proper, meaning that it is equally sensitive to sharpness (eq. 4.4) and reliability (eq. 4.5). A maximized skill score (i.e., $I_S^\alpha = 0$) corresponds to an optimal solution.

$$I_S^\alpha = \sum_{(I^\alpha, E_A) \in I_{Test}} \begin{cases} -2\alpha I_U^\alpha - 4(E_{P,Min} - E_A) & \text{if } E_A < E_{P,Min} \\ -2\alpha I_U^\alpha & \text{if } E_A \in I^\alpha \\ -2\alpha I_U^\alpha - 4(E_A - E_{P,Max}) & \text{if } E_A > E_{P,Max}. \end{cases} \quad (4.6)$$

Prediction intervals should not be confused with confidence intervals. Confidence intervals show the uncertainty of the model estimate but they do not show the estimated range of values for a specific new observation. Confidence intervals only show a range of possible values for the model parameters. Prediction intervals are always wider than confidence intervals, because their calculation takes into account wrong model parameters (just like confidence intervals) and also the spread of the data (e.g., the monthly prediction error).

Although a prediction interval provides less information than a full probability density function, it gives a robust estimate of the uncertainty of the prediction for a variable, because it can be calculated with quantile regression which does not make assumptions about the parametric form of $p(E_A)$ (sec. 4.3.1).

4.3.4. Quantile regression

Prediction intervals can also be calculated with quantile regression (QR). Quantile regression is used by (Zarnani & Musilek, 2013; Bremnes, 2004; Nielsen et al., 2006; Pinson et al., 2007a) to estimate uncertainty in weather and weather-dependent energy predictions (power output of wind turbines). Like probability density estimation it estimates uncertainty in predictions but it is independent from distributional assumptions about the data.

Variables (i.e., regressors) that contain information about prediction uncertainty can easily be added to the quantile regression model. This enables a systematic search for new regressors that improve the skill score (eq. 4.6) of the prediction intervals. For these reasons quantile regression is selected as method to estimate uncertainty in energy predictions in this work. The quantile regression model is described in chapter 5.

5. The Quantile Regression Model

Abstract: *In chapter 4 quantile regression (QR) was selected as a method to estimate the uncertainty of predictions with prediction intervals. This chapter describes the quantile regression model. Furthermore, methods are described to reduce a weather prediction to a small set of variables to be used as regressors in the QR model (in addition to the variables which have been identified in sec. 3.3.5). In this context the North Atlantic Oscillation index (NAOI)—which summarizes the air pressure difference in the North Atlantic Ocean—is introduced in this chapter. Finally, two new regressors (Local Energy Distribution Moments (LEDM)) are introduced which are calculated by the mean and variance of the predicted energy in a region of interest. At the end of the chapter a short discussion compares the NAOI with the LEDM.*

Contents

5.1. Linear quantile regression (QR) 57
5.2. Regressors for the QR model 59
    5.2.1. Principal component analysis 60
    5.2.2. North Atlantic Oscillation Index (NAOI) 61
    5.2.3. Other climatological indices 62
    5.2.4. Statistical moments 63
    5.2.5. Local Energy Distribution Moments (LEDM) 65
    5.2.6. Relation between LEDM and NAOI 69

5.1. Linear quantile regression (QR)

The quantile regression model calculates conditional quantiles Q (or more precisely percentiles) of the wind energy which is calculated with historical weather data. For example, the most well-known quantile is $Q(0.5)$ which is the median. The quantiles are calculated conditionally on the values in a vector of regressors $\mathbf{x} \in \mathbb{R}^q$, $q > 0$. $\mathbf{x}$ contains variables which are related to the dependent variable E_A. The bold notation generally indicates that a variable is a vector. A vector with a subscript, e.g., x_i, denotes the ith element of $\mathbf{x}$. A pair of quantiles

$$I^\alpha = [Q(\alpha_1), Q(\alpha_2)], \alpha_1 \in [0,1], \alpha_2 \in [0,1] \quad (5.1)$$

with $\alpha_1 < \alpha_2$ constitutes a prediction interval for E_A with a probability of $100 \cdot (\alpha_2 - \alpha_1)$ % that $E_A \in I^\alpha$. The sample quantile $Q(\alpha)$ (eq. 5.2 and fig. 5.1) is modeled as a linear combination of the known regressor vector **x** and the unknown weight vector β (eq. 5.2).

$$Q(\alpha) = \beta_0(\alpha) + \beta_1(\alpha)x_1 + \ldots + \beta_q(\alpha)x_q \qquad (5.2)$$

Figure 5.1: *The integral of the distribution between $Q(\alpha_1)$ and $Q(\alpha_1)$ shows the spread of a data sample (for example historical prediction errors). If, for example, $\alpha=20$, than the integral of the function between $Q(\alpha_1)$ and $Q(\alpha_1)$ contains 80 % of the distribution.*

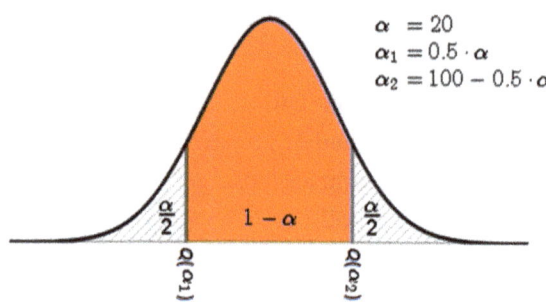

Given a sample of historical data $H = \{(\mathbf{x}, E_A)_i\}_{i=1}^{N}$, the weight vector β is obtained by linear optimization. For this purpose the model error for a data observation is defined by $m_{err} = E_A - (\beta_0 + \beta_1 x_1 + \ldots + \beta_q x_q)$ which is the difference between the observed wind energy E_A in the data sample and the corresponding model prediction based on the values in **x**. The weight vector β is obtained by minimizing (eq. 5.3) for the data set H

$$\hat{\beta}(\alpha) = \arg\min_{\beta} \sum_{i=1}^{N} \rho_\alpha(m_{err}) \qquad (5.3)$$

where ρ_α is the loss function in eq. 5.4. It is also called the tilted absolute value function or check function in fig. 5.2 by (Koenker & Bassett, 1978).

$$\rho_\alpha(m_{err}) = \begin{cases} m_{err} \cdot \alpha & \text{if } m_{err} \geq \alpha \\ m_{err} \cdot (\alpha-1) & \text{if } m_{err} < \alpha. \end{cases} \qquad (5.4)$$

It puts asymmetrical weights on the residuals so that eq. 5.2 yields the desired quantiles. It is noted, that if eq. 5.4 is replaced with $\rho_\alpha(m_{err}) = m_{err}^2$, than eq. 5.2 leads to ordinary least-square estimates instead of quantiles (Pinson et al., 2007a). This is a linear optimization problem which can be solved with the **GNU R quantreg** software package by Koenker (2015).

Figure 5.2: *The check function with α = 0.25. Data points are divided into three sets based on their associated residuals $y_i - f(x_i)$. The three sets are called left, elbow, and right (figure and text by (Li & Zhu, 2008, p. 2)).*

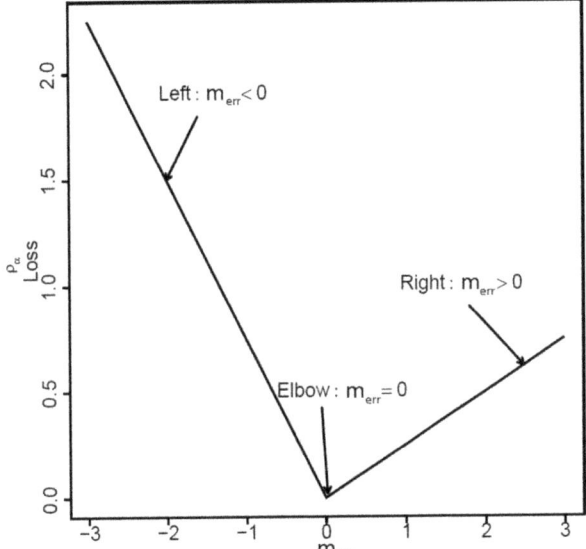

5.2. Regressors for the QR model

The accuracy of meteorological predictions varies depending on the general weather situation. Estimating the uncertainty in a prediction requires to gather information that indicates the level of uncertainty. For this purpose the available data need to be reduced to variables which contain the relevant information. Depending on the context, these variables are called predictors (in statistics), explanatory variables or risk indices (in forecasting) or conditional variables (in stochastics). The term *regressor* is the mathematical term and it is used here. A well-known method for data dimension reduction is principal component analysis (PCA).

A similar method in meteorology is the empirical orthogonality function (EOF). The North Atlantic Oscillation index (NAOI) for example is calculated with EOF. Sections 5.2.1 and 5.2.2 discuss PCA in general and the NAOI as a specific implementation to extract information about prediction uncertainty from weather data.

5.2.1. Principal component analysis

Principal component analysis (PCA) describes a data matrix by a linear combination of the eigenvectors of the covariance matrix of the original data set. These eigenvectors are the principal components. Each eigenvector describes the variability of the data in one dimension. The method is implemented iteratively by fitting orthogonal lines through the data starting with the direction of highest data variance as depicted in fig. 5.3.

PCA is used for example in data compression but it is also applied in synoptic climatology. By discarding eigenvectors which explain little of the variability in the data it is possible to remove noise from data and to identify typical macro weather situations. PCA is used by Pinson *et al.* (2007a) to identify typical weather situations in the German North Sea. With a combination of PCA, hierarchical clustering and kernel density estimation they estimate prediction uncertainty for different macro weather situations.

Their approach is based on multiple weather parameters (humidity, temperature, air pressure) that are not available here. As a result they observe significant variability in prediction uncertainty for different macro weather situations. They confirm that dynamic low pressure situations with fronts are related to considerably larger (factor 1.5–1.7) prediction errors than weather situations that are mainly influenced by rather stationary high pressure systems. They also point out that spatial pressure gradients are of great significance for weather classification.

Here PCA is not applied and will not be discussed further because of the reasons below. A more complete introduction to Principal Component Analysis is given by Jolliffe (2002).

1. According to (Khan, 1998, p. 62) weather is defined as the state of the atmosphere as determined by meteorological phenomena including temperature, precipitation, winds, clouds, sunshine, pressure and visibility. PCA aims to classify weather situations, therefore important weather variables (sea & air temperature, pressure gradients, etc.) are required but not available here.
2. The PCA is only capable of identifying linear relationships, but the relation between weather and application-specific energy models is often non-linear (e.g., wave height versus wave resistance on a ship). The radial wind and wave direction data is also an issue because PCA is based on the euclidean distance between data points.

3. PCA and clustering require careful parameter tuning for individual sites (e.g., the number of clusters in agglomerative or hierarchical clustering), thus it is not suitable for non-stationary applications such as weather routing (see the comparison between weather routing and wind power generation in sec. 2.3).

Figure 5.3: Principal component analysis after two iterations. The eigenvectors v1 and v2 point into the direction with the highest variance in the data. Image: (Jehan, 2005, p. 83).

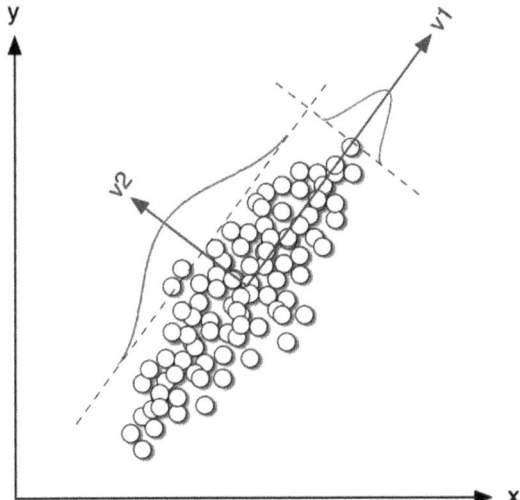

5.2.2. North Atlantic Oscillation Index (NAOI)

The North Atlantic Oscillation index (NAOI) represents the state of atmospheric circulation in the North Atlantic. Traditionally, the NAOI was computed from the pressure gradient between the Azores high and the Icelandic low in fig. 5.4. Today it is calculated with Empirical Orthogonal Function analysis from 2-dimensional air pressure data (Hurrell *et al.*, 2003) from sea level. The North Atlantic Oscillation influences the direction of general storm paths for major North Atlantic tropical cyclones. A northern position of the Azores high allows storms to track up the North American coast (Scott *et al.*, 2003). Therefore, the daily NAOI potentially contains information about the uncertainty of predictions.

Figure 5.4: The North Atlantic Oscillation pattern[6].

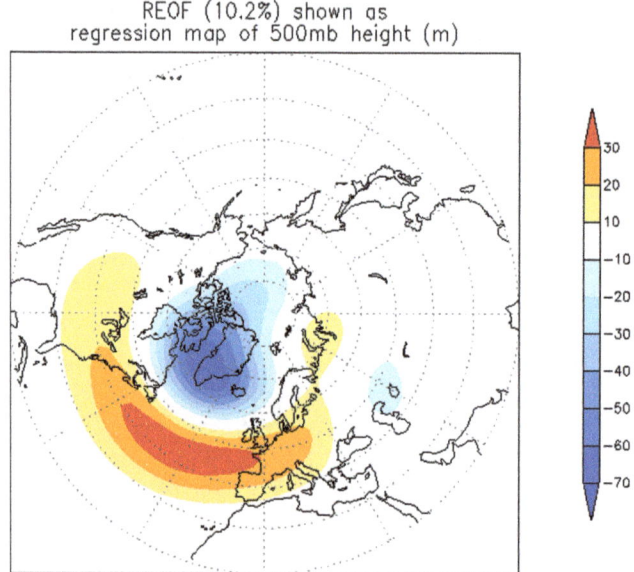

5.2.3. Other climatological indices

The Ocean Observations Panel for Climate (OOPC) has launched the "State of the Oceans" project[7] to describe the state of the ocean by a few climate index time series that are calculated from sea surface temperatures and air pressure anomalies. Each index is calculated as a real number in combination with an expected variance. Thus it can be used as input variable to a stochastic model to compute the uncertainty induced by the weather conditions. Additionally, the American National Climatic Data Center (NCDC) provides a selection of time series for various climatic indices. Yet the NAOI is the only index (that is also updated on a daily basis) for which a relation to storm tracks and intensities in the North Atlantic Ocean is documented in the literature.

---

6   http://www.cpc.ncep.noaa.gov/products/precip/CWlink/pna/new.nao.loading.gif (accessed 30.12.2015).

7   http://stateoftheocean.osmc.noaa.gov/ (accessed 08.07.2015).

5.2.4. Statistical moments

As an alternative to the existing (climatological) indices a new index is defined here that summarizes the weather situation. For this purpose it is assumed that the weather is a random variable that is completely defined by its probability density function. Since the distribution type is unknown it has to be described by estimates of the distribution's moments. The moments are estimated from a data sample of observations $\{o_j\}_{j=1}^N$ to describe the properties of the distribution. Statistical moments are the sums of integer powers of the values in the data sample. The first moment is the mean in eq. 5.5.

$$\bar{o} = \frac{1}{N} \sum_{j=1}^{N} o_j \quad (5.5)$$

Alternatively, the median (eq. 5.6) or the mode are estimators for the central value of the distribution. The median is the value for which larger and smaller values have the same probability. It can be found by sorting the values and finding the value which has equal numbers of values above and below it. This is a process of order $N \cdot log(N)$. However, there are more efficient but less intuitive algorithms for this which run in linear time, e.g., by Hoare (1961). In case of an equal number the median is defined as the mean of the two central values. The mode is defined as the maximum value of the distribution sample. It is useful to find the central value if there is a single sharp maximum.

Modes can be useful if a distribution is multi-modal. In those cases the mean and median fail to identify the central value. Both median and mode are robust against outliers, but require a sufficient large data sample. It is assumed that there are no outliers in the input data for the predictions (i.e., in the weather prediction), hence the mean is used here instead.

$$o_{med} = \begin{cases} o_{(N+1)/2} & \text{if N is odd} \\ \frac{1}{2}(o_{N/2} + o_{(N/2)+1}) & \text{if N is even} \end{cases} \quad (5.6)$$

The second moment in eq. 5.7 is the variance (mean squared deviation of o_j from $\bar{o}$) which shows the spread of the values around the central value.

$$Var(o_1, \ldots, o_N) = \frac{1}{N-1} \sum_{j=1}^{N} (o_j - \bar{o})^2 \quad (5.7)$$

Here the corrected two-pass algorithm of (Chan et al., 1983) is used to calculate the variance, because it corrects round-off errors. The second term is an approximation of the rounding error of the first term. In other words the rounding errors of the first term are canceled out by the rounding errors of the second term. The second term is usually much smaller.

$$Var_{corr}(o_1,\ldots,o_N) = \frac{1}{N-1}\left\{\sum_{j=1}^{N}(o_j-\overline{o})^2 - \frac{1}{N}\left[\sum_{j=1}^{N}(o_j-\overline{o})\right]^2\right\} \quad (5.8)$$

If the mean was known a priori the denominator would be N instead of $N - 1$ in eq. 5.8. The difference is irrelevant if the sample size is not too small. For some distribution types, statistical moments do not exist (e.g., Cauchy distribution), i.e., the mean and variance do not converge with an increasing number of points and different samples will show no consistency even if they are drawn from the same distribution. In those cases eq. 5.5 and eq. 5.8 do not estimate the moments of the unknown distribution. The average deviation or mean absolute deviation *ADev* in eq. 5.9 is a more robust estimator for the spread of the data.

$$ADev(o_1,\ldots,o_N) = \frac{1}{N}\sum_{j=1}^{N}|o_j-\overline{o}| \quad (5.9)$$

Often $\overline{o}$ is replaced by the sample median, since it has been proven to minimize *ADev* for any fixed sample.

Kurtosis and Skewness The interpretation of higher moments of the data depends on the modality of the unknown distribution. If the distribution is multi-modal the meaning of the values of higher moments is completely different, hence they are not suitable to describe an unknown probability distribution. The mean and the variance are the first and second moments of any statistical distribution.

The skewness is the third moment, and the kurtosis is the fourth moment. Skewness measures asymmetry of the distribution and it is non-dimensional. The mean and standard deviation are dimensional values, i.e., they have the same units as the original data. Zero skewness denotes a perfectly symmetrical distribution. Skewness implicitly assumes an uni-modal distribution and depends on the tails of the distribution.

Kurtosis is also a non-dimensional quantity which measures the relative peakedness (or flatness) of a distribution relative to a normal distribution. Since higher moments are less stable and implicitly refer to an uni-modal (normal) distribution they are not used here to describe the distribution. Table 5.1 summarizes the variables which are available to estimate moments of the unknown distribution. The median, mode and average deviation are robust estimators which are even applicable if the distribution has no moments.

5.2. Regressors for the QR model

Table 5.1: Estimators for statistical distribution moments

variable	name	distribution moment	properties
$\bar{o}$	mean	1st (center)	insensitive to outliers or
σ^2	variance	2nd (spread)	extreme values
o_{max}	mode[8]	1st (center)	unsuitable for multi-modal distributions
o_{med}	median[8]	1st (center)	slow adaptation to broad
o_{mae}	avg. deviation[8]	2nd (spread)	distribution tails

Yet there is no reason to assume such a distribution here since the number of different possible weather scenarios is finite. Hence the mean and variance are used, higher moments are not considered since they imply an uni-modal distribution.

5.2.5. Local Energy Distribution Moments (LEDM)

The North Atlantic Oscillation index (NAOI) describes the macro weather situation in the North Atlantic ocean by a single number, therefore, depending on the application, this index might be too general to uniquely identify specific weather situations. In addition, the NAOI is calculated from air pressure data and the linkage between air pressure, wind and the application-specific energy model is complex. Therefore, another index is calculated here that is based on statistical moments of the energy distribution in the local region of interest.

5.2.5.1. Energy distribution in the region of interest

Quantile regression is based on a vector of regressors **x** which contains information about the uncertainty of the prediction. Some regressor variables were proposed, e.g., the meteo-risk index (MRI) by Pinson & Kariniotakis (2004). Yet a study by Nielsen *et al.* (2006) showed that prediction intervals for wind power generation were not significantly improved by regressor variables which are calculated from single weather parameters, e.g., mean regional wind speeds.

Consequently, the prediction intervals in this work are based on regressors which are calculated from the distribution of the predicted wind energy in the region of interest **roi**. The region of interest with respect to the complete route consists of multiple smaller regions of interest that are centered on the waypoints

---

8 robust against *outliers*.

of the route. The region of interest of the *l*-th waypoint of a route with n waypoints (eq. 5.10) is a vector containing the k nearest coordinates[9] in the gridded weather data with respect to pos_l in fig. 5.5.

$$roi(pos_l) = (roi_{1,l}, \ldots, roi_{i,l}, \ldots, roi_{k,l}) \qquad (5.10)$$

The notation $w_d^h(p)$ denotes the predicted wind and wave data at position p at date d with a forecast horizon h.

$$w_d^h = \begin{pmatrix} wi_s \\ wi_d \\ swh \\ wa_d \\ wa_p \end{pmatrix} \quad \begin{array}{l} wi_s = \text{average wind speed 10 m above sea level} \\ wi_d = \text{average wind direction 10 m above sea level} \\ swh = \text{significant wave height} \\ wa_d = \text{mean direction of waves} \\ wa_p = \text{mean wave period} \end{array} \qquad (5.11)$$

The weather in the region of interest of the complete route is then defined by eq. 5.12, where Δt is the travel time between two waypoints. It should be noted that this notation implies a ship with a constant speed.

$$W_{roi} = \begin{bmatrix} w_d^{0 \cdot \Delta t}(roi_{1,1}) & \cdots & w_d^{n \cdot \Delta t}(roi_{k,1}) \\ \vdots & \ddots & \vdots \\ w_d^{0 \cdot \Delta t}(roi_{1,n}) & \cdots & w_d^{n \cdot \Delta t}(roi_{k,n}) \end{bmatrix} \qquad (5.12)$$

The region of interest for a single waypoint of a ship route is depicted in fig. 5.5 (circular area).

---

9 By the spherical law of cosines of (Abramowitz & Stegun, 1972, p. 79) the distance between two coordinates on a sphere is the arc length for a sphere of radius $r(=6371$ km).

Figure 5.5: *The region of interest **roi** covers the k nearest grid cells in the GSM model. The weather from the region of interest **roi** (circular area) is converted into (potential) propulsion energy. The propulsion energy is calculated with the actual ship heading at the central point pos_i, since it is assumed that the ship sticks with its original course but the weather might shift unexpectedly. An optimal value for k is determined by alg. 1.*

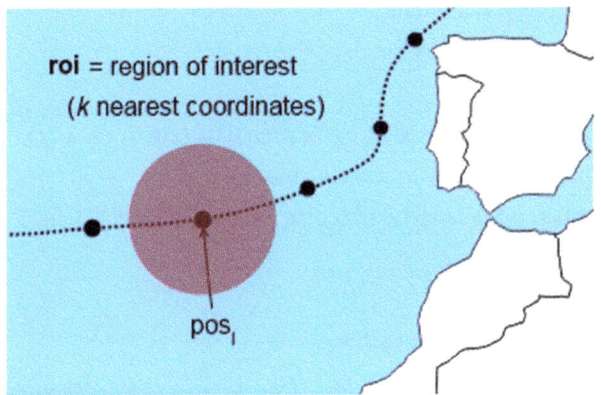

5.2.5.2. Size of the region of interest

The size k of **roi** is determined by cross-validation with the historical data (alg. 1). The algorithm calculates prediction intervals and skill scores for historical weather predictions and iterates over k. Each year of historical data is used once with each value of k to validate the model *(cross validation)*, except for the year that is evaluated. For example, the data from 2005 to 2009 (i.e., ≈ 1500 scenarios) are used in order to determine an optimal value of k for 2010. Afterwards, the value of k that yields the highest average skill score, is returned.

Algorithm 1: Cross-validation to determine optimal size k of region of interest.

    input   : historical weather data
    output : Optimal size k of region of interest.
    for each *year* $y \in$ *historical data* **do**
        set data from y as test data
        set other data as calibration data
        for $k \in [k_{min}, k_{max}]$ **do**
            calculate regressors **x** for calibration data (alg. 2, eq. 2.21)
            calculate intervals for test data (eq. 5.2)
            calculate normalized skill scores for test data (eq. 4.6, fig. 7.2)
        end
    end
    calculate average scores for each k
    return k with highest scores

Calculating the mean $E_{P,M}$ and variance $E_{P,V}$ of the energy distribution in the regions of interest of all waypoints (alg. 2) yields two regressors for the quantile regression model. Another regressor is the prediction E_P itself which indicates the center of the prediction interval. Since the mean and the variance represent statistical moments of the energy distribution they are henceforth referred to as Local Energy Distribution Moments (LEDM).

Algorithm 2: Calculation of Local Energy Distribution Moments.

    input   : region of interest **roi**, k
    output : vector of regressors x

$$E_{P,M}(k) = \frac{1}{n \cdot k} \cdot \sum_{w_d^h \in W_{roi}} E_P(w_d^h)$$

$$E_{P,V}(k) = \sqrt{\frac{1}{n \cdot k} \cdot \sum_{w_d^h \in W_{roi}} (E_P(w_d^h) - E_{P,M})^2}$$

$$\mathbf{x} = (E_{P,M} \quad E_{P,V})^T$$

Several regressors for the quantile regression model have already been identified in tbl. 3.4 in sec. 2.3. The predicted energy E_P is expected to be the most significant indicator for E_A. Now the table is extended with the NAOI and the LEDM regressors. The corresponding quantile regression (QR) models are shown in tbl. 5.2. The prediction intervals that are calculated with these models are compared in chap. 7 to identify the combination of regressors which yields the highest skill score.

Table 5.2: *Quantile regression models, the model results are compared in chap. 7.*

model name	predictors
QR_E	E_P
QR_m	E_P, month
QR_h	E_P, hor_j
QR_{LEDM}	E_P, $E_{P,M}$, $E_{P,V}$
QR_{NAOI}	E_P, NAOI

$\rightarrow E_A$

With the definition of the predictors (i.e., regressors) the quantile regression model is complete. The model calculates the uncertainty of predictions for the complete route. The complete dataset (except for the data that are evaluated) is used to calibrate the model. The values E_P, $E_{P,M}$ and $E_{P,V}$ determine the widths of the predictions intervals.

5.2.6. Relation between LEDM and NAOI

The North Atlantic Oscillation index (NAOI) describes the pressure gradient between the Azores High and the Icelandic Low. The NAOI is traditionally used by meteorologists to analyze and predict the weather in the North Atlantic Ocean. Studies by (Rogers, 1997; Walter & Graf, 2005) confirmed a statistical correlation between the tracks and intensities of depressions in the North Atlantic Ocean. For this reason the NAOI is a potential indicator for prediction uncertainty—depending on the specific application.

For example, predictions for the power output of wind turbines mostly depend on the wind speed whereas predictions for the propulsion power of a ship also depend on the wind direction and wave conditions (sec. 2.3). In contrast to that, Local Energy Distribution Moments (LEDM) are calculated from the predicted energy (whereas the NAOI is calculated from the weather analysis), hence the relation between weather and energy is already taken into account. This is important for applications where the relation between weather parameters and energy is complex, e.g., weather routing.

For example, the effect of wind on the ship is ambiguously—depending on the wind direction. This shows, that information about the uncertainty of a prediction for some weather-dependent variable (e.g., wind energy) is easier to interpret if it is calculated based on the variable itself instead of individual weather parameters that have an impact on the value of the variable.

6. Implementation

Abstract: *Chapter 5 introduced quantile regression and new regressors for a linear quantile regression model to estimate the uncertainty of weather-dependent energy predictions with prediction intervals. This chapter describes the implementation of the quantile model in* GNU R *and the* Java *implementation of a A* weather routing application. The integration of the prediction intervals into the weather routing application is also described.*

Contents

6.1. Database with historical weather predictions 71
6.2. A* route optimization 72
6.3. Route optimization with uncertainty 75
6.4. Quantile regression 76

6.1. Database with historical weather predictions

The weather predictions and analyses for the time period between October 2005 and November 2010 were provided as Zip-archives by the German Weather Service. A batch script is used to extract the data and write it into a MySQL database using the LOAD DATA INFILE command.

The database consists of 59 tables with ≈ 380 GB of data. 58 tables contain global historical prediction data (3 hours of prediction per table; $3 \cdot 58 = 174$ hours) and one additional table contains the historical weather analysis data. Although the LOAD DATA INFILE command is very efficient compared with a row-wise insertion of the data the parsing of the data takes more than 48 hours. The database access is encapsulated by a Java interface based on the JDBC driver to access the weather data from the route optimization simulation. Experiments showed that the database access is essential for the performance of the simulation. Therefore, the runtime of the simulation and the runtime of the uncertainty estimates depend on the efficiency of the implementation of these modules. The UML class diagram in appendix C.6 shows the Java code structure.

The NWP data are defined in WeatherDBI by the weather model resolution, the different weather types (wind speed, wave height, etc.) and the order in which the data are stored in the database. Based on this information, weather data for a given date, prediction horizon, location and type are read from the database. The simulation can be run with either weather predictions or weather analyses by instantiating either an object of class `DWDForecast` or `DWDAnalysis` for the abstract interface `WeatherDBI`. Independently from the specific database in use the class `DBCache` assures that weather data are not read twice from the database during a single simulation run. In addition, if weather from a grid coordinate is loaded into the application, the surrounding weather within a radius of `lon_cache_size` · `lat_cache_size` is loaded into the cache, too.

The UML diagrams which describe the Java implementation of the weather routing algorithm and the ship propulsion model are depicted in appendix C.

North Atlantic Oscillation index (NAOI) The values of the historical daily NAO index since January 1950 were downloaded from an FTP server of the American Climate Prediction Center[10]. The values were computed by empirical orthogonal functions from oceanic weather analyses.

6.2. A* route optimization

The A^* search algorithm is an extension of Dijkstra's algorithm. It operates on a tree (i.e., a graph) and uses a heuristic to remove paths from the graph that are longer than other known paths early. To this end each new waypoint *neighbor* is ranked by E_T (eq. 6.3) where $E_1(neighbor)$ (eq. 6.1) is the required ship propulsion energy from the departure to the waypoint *neighbor* on the path in the tree and $E_2(neighbor)$ is the predicted ship propulsion energy $E_2(neighbor)$ (eq. 6.2) on the orthodrome from *neighbor* to the destination.

$$E_1\ (neighbor) = E_P\ (departure, neighbor) \tag{6.1}$$

$$E_2\ (neighbor) = E_P\ (neighbor, destination) \cdot 0.7 \tag{6.2}$$

$$E_T\ (neighbor) = E_1\ (neighbor) + E_2(neighbor) \tag{6.2}$$

In each step the A^* algorithm (alg. 3) selects the waypoint with the lowest rank $E_T\ (neighbor)$ that has not been expanded in the tree and calculates a set of neighbors based on a predefined ship speed V_g, the travel time per step Δt and the angle between candidate ship headings $\Delta\phi$ as depicted in fig. 6.1.

The algorithm terminates when the lowest-ranked waypoint equals the destination. If $h(x)$ is an admissible heuristic (i.e., it does not overestimate the true propulsion energy to get to the destination), A^* has been proven to be both optimal and complete. For this reason, eq. 6.2 is scaled by a factor of 0.7 to ensure that E_2 does not overestimate the actual propulsion energy. Thus, if a path to the destination exists it will be found and it will be optimal in terms of minimal ship

---

10 Available from ftp://ftp.cpc.ncep.noaa.gov/cwlinks/ (accessed 26.11.2015).

propulsion energy. But since the weather prediction can be inaccurate there is no guarantee that the optimized path is actually optimal.

Algorithm 3: Pseudocode of A^* search algorithm (adapted from (Jones, 2015, pp. 57–58)). The relation between t_{max} and E_A is depicted in fig 7.18.

input : *departure, destination*, number of neighbors *kNN*, angle between segments $\Delta\phi$, travel time per segment Δt, maximum allowed travel time t_{max}, ship speed V_s, heuristic E_2

output : optimized *route* from *departure* to *destination*

OPEN = *priority queue containing departure*

while *lowest rank in OPEN not equals destination* **do**
    next = remove lowest rank item from OPEN
    for each *neighbor of next* **do**
        $E_{new} = E_1(next) + E_P(next, neighbor)$
        if *neighbor in OPEN and* $E_{new} < E_1(neighbor)$ **then**
            remove neighbor from OPEN, because new path is better
        end
        if $t(departure, neighbor) > t_{max}$ **then**
            remove neighbor from OPEN, because t_{max} exceeded
        end
        if *neighbor **not** in OPEN* **then**
            set $E_1(neighbor)$ to E_{new}
            add neighbor to OPEN
            set priority queue rank to $E_1(neighbor) + E_2(neighbor)$
            set neighbor's parent to next (fig. 6.1)
        end
    end
end
return *reverse path from goal to start by following parent pointers*

Figure 6.1: A* search algorithm after 2 iteration steps (kNN=opening of tree). In each step the waypoint with the lowest E_T value is selected and expanded with neighbors. One neighbor lies on the orthodrome between the last waypoint and the destination, the other neighbors are arranged symmetrically at each side of the orthodrome. Each new neighbor is stored in a priority queue OPEN. Thus, the algorithm successively builds up a tree of partial paths. The algorithm is described in alg. 3.

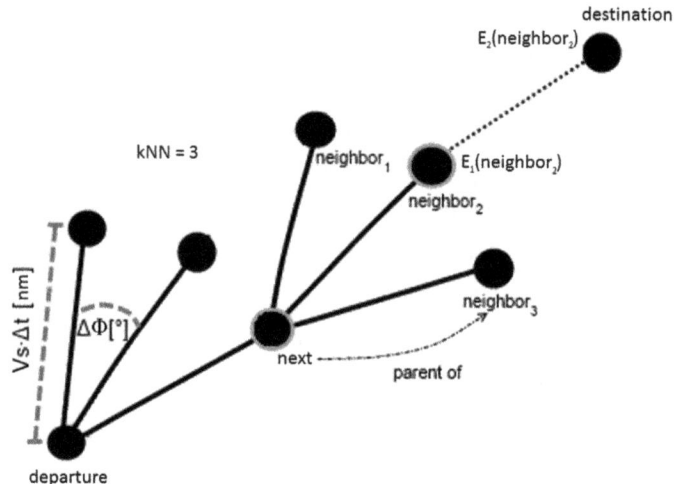

6.3. Route optimization with uncertainty

*Figure 6.2: Image shows the circular regions of interest (**roi**) of the route environment.[11] The distribution of energy in **roi** indicates uncertainty ($E_{P,M}$, $E_{P,V}$ in alg. 2) of predictions for the (wind-assisted) ship propulsion.*

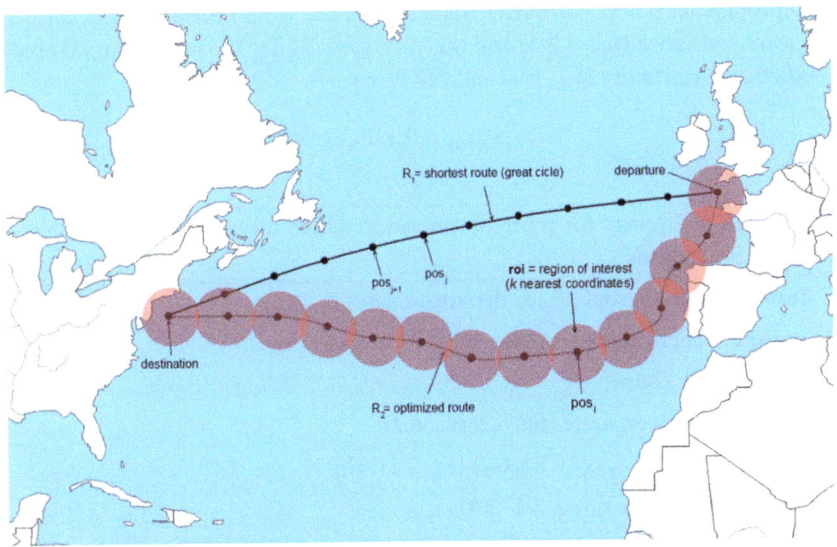

Figure 6.2 shows the shortest route R_1 and an optimized shipping route R_2 from Europe to North America. The route has been optimized with a branch-and-bound weather routing algorithm for minimization of the ship energy consumption that is described in sec. 6.2. Two strategies for weather routing are compared here.

The first strategy optimizes the route in fig. 6.2 with historical weather predictions from 2008 once per week. Then the true propulsion energy E_A and the prediction error (eq. 4.3) are calculated using historical weather analyses.

The second strategy in alg. 4 additionally takes into account prediction uncertainty. Here the optimized route is selected only if the uncertainty I_U^α (eq. 4.4) does not exceed a predefined threshold $I_{U,Max}^\alpha$. Otherwise the great circle R_1 is selected as traveling route. Thus at departure time (and only then) it is determined if the ship travels on an optimized route. After that there is no fallback to the great circle but the shipping route is continuously re-optimized instead, until the ship arrives at its destination.

---

11 http://www.d-maps.com/m/world/atlantiquenord/atlantiquenord05.gif (modified, accessed 31.12.2015).

The optimal size k of the region of interest **roi** is determined via crossvalidation of the normalized skill score (eq. 4.6) with the historical weather data (sec. 2.4.1.1). In both strategies the ship motor power $E_M(v)$ is set to 4000 kW (bulker carrier at a constant speed of 13 kn). Therefore, the variability of the ship propulsion power completely depends on the wind resistance (sec. 2.2.3), the wave resistance (sec. 2.2.4) and the WPS (sec. 2.2.1). The maximum allowed prediction uncertainty $I^\alpha_{U,Max}$ is calculated by eq. 6.4.

$$I^\alpha_{U,Max} = 0.5 \cdot E_P \quad (6.4)$$

Algorithm 4: Strategy for route optimization which takes into account uncertainty in predictions.

input : shortest route R_1, k, departure dates D, $I^\alpha_{U,Max}$
output : propulsion energy E_D, prediction error Err_D
for each $d \in D$ **do**
　　R_2 = optimized route (fig. 6.2, sec. 6.2)
　　$\mathbf{x} = (E_P(d, R_2), E_{P,M}(k), E_{P,V}(k))$ (eq. 2.21, alg. 2)
　　$I^\alpha = [Q(\alpha_1), Q(\alpha_2)]$ (eq. 5.1-5.4)
　　if $I^\alpha_U > I^\alpha_{U,Max}$ **then**
　　　　add $E_{P,Err}(d, R_1)$ to $\mathbf{Err_D}$ (eq. 2.23)
　　　　add $E_P(d, R_1)$ to $\mathbf{E_D}$ (eq. 2.21)
　　else
　　　　add $E_{P,Err}(d, R_2)$ to $\mathbf{Err_D}$ (eq. 2.23)
　　　　add $E_P(d, R_2)$ to $\mathbf{E_D}$ (eq. 2.21)
　　end
end

6.4. Quantile regression

Quantile regression was implemented with the GNU R `quantreg` software package by Koenker (2015). The software is freely available and has been applied in multiple other related applications to calculate prediction intervals. The code and comments below were adapted from code by Pinson *et al.* (2007a). The example shows the quantile regression function with three regressors $\mathbf{x} = (E_P, E_{P,M}, E_{P,V})$. The data are split into `train` and `test` data sets

6.4. Quantile regression

to calibrate and validate the QR_{LEDM} model. The function also estimates conditional quantiles for models with a different set of regressors, which requires only minor code modifications.

```
QuantileRegression <- function(historicalData, testYear=2007,
                            x1='EP', x2='EPM', x3='EPV',
                            y ='EA', alpha=0.05)
{
    dat = read.csv(historicalData, dec=".");
    dat$X <- NULL
    dat = dat[complete.cases(dat),];

    ## Comment: Splitting dataset into 'train' and 'test' data

    train = subset(dat[,],dat[,2] != testYear);
    test  = subset(dat[,],dat[,2] == testYear);

    cols <- c(x1, x2, x3, y);

    train = train[,cols, drop=FALSE];
    test  = test[,cols, drop=FALSE];

    train = subset(train[,], train[,1]>0);
    test  = subset(test[,], test[,1]>0);

    names(train) <- c("x1", "x2", "x3", "y")
    names(test)  <- c("x1", "x2", "x3", "yt")

    ## Comment: Make natural spline bases with 1 column (i.e., linear)
    ## and knots placed according to the quantiles of x1, x2 and x3:

    basis.x1<-ns(train$x1,df=1,intercept=F)
    basis.x2<-ns(train$x2,df=1,intercept=F)
    basis.x3<-ns(train$x3,df=1,intercept=F)

    ## Fitting lower and upper quantile models (1 denotes the intercept):

    qUp = 1 - alpha/2;
    qLo = alpha/2;

    loFit<-rq(train$y~1+basis.x1+basis.x2+basis.x3,tau=qLo)
    upFit<-rq(train$y~1+basis.x1+basis.x2+basis.x3,tau=qUp)

    coef(loFit)
    coef(upFit)

    ## Comment: Calculating prediction intervals for data frame 'test':

    test.b.x1<-ns(test$x1,
        knots=attributes(basis.x1)$knots,
        Boundary.knots=attributes(basis.x1)$Boundary.knots,
        intercept=attributes(basis.x1)$intercept)
```

```
    test.b.x2<-ns(test$x2,
       knots=attributes(basis.x2)$knots,
       Boundary.knots=attributes(basis.x2)$Boundary.knots,
       intercept=attributes(basis.x2)$intercept)

    test.b.x3<-ns(test$x3,
       knots=attributes(basis.x3)$knots,
       Boundary.knots=attributes(basis.x3)$Boundary.knots,
       intercept=attributes(basis.x3)$intercept)

    ## Comment: The forecast is constructed for the 'test' dataset.
    ## It is important that the coefficients above are estimated
    ## based on a different 'train' dataset as outlined above.

    qForecast<-data.frame(
      Qlo=cbind(1,test.b.x1,test.b.x2,test.b.x3)%*%coef(loFit),
          Qup=cbind(1,test.b.x1,test.b.x2,test.b.x3)%*%coef(upFit))

    return(qForecast);
}
```

7. Evaluation and Results

Abstract: *This chapter describes the prediction intervals calculated with quantile regression for the ship propulsion energy and the travel time on the route in fig. 6.2. The prediction intervals are evaluated with respect to sharpness, reliability and the skill score. The key aspects of the evaluation are the accurateness of uncertainty estimates with the new regressors (LEDM), the impact of a WPS on prediction uncertainty and the integration of uncertainty estimates into an existing weather routing software. It is demonstrated that the prediction accuracy and the average ship propulsion energy significantly improve if the uncertainty of predictions is taken into account in the weather routing strategy.*

Contents

7.1. Evaluation of prediction intervals 79
7.2. Application to wind-assisted sailing propulsion 80
7.3. Prediction intervals for ship propulsion energy 83
7.4. Prediction intervals for travel time 86
7.5. Uncertainty in weather routing with WaSP 88
7.6. Prediction interval benchmarks 89
7.7. Runtime discussion 92
7.8. Annual cost savings for a multi-purpose carrier 92

7.1. Evaluation of prediction intervals

Local Energy Distribution Moments (LEDM) were proposed as regressors for the quantile regression model in chap. 5. As a benchmark multiple alternative regressors are defined in tbl. 7.1. The basic QR_E model uses only the predicted energy consumption E_P as regressor. The QR_{NAOI} model uses the daily NAO index $NAOI \in [-3, 3]$ (sec. 5.2.2) and E_P as regressors. The QR_m model uses the current month $\in [1,12]$, i.e., a purely statistical indicator and E_P. The QR_{LEDM} model uses E_P and the mean $E_{P,M}$ and variance $E_{P,V}$ of the energy distribution in the region of interest (described in sec. 5.2.5.1).

The QR_h model from tbl. 5.2 is not used here since the model is only applied to one route, i.e., hor_j is constant. The QR_T is used to calculate prediction intervals for the travel time T_A, $\mathbf{x}$ only contains the predicted travel time T_P. This shows that the quantile regression model can also be used to calculate the uncertainty for other variables than the propulsion energy of the ship.

Table 7.1: *Quantile regression models for energy and time prediction*

$$
\begin{array}{ll}
QR_E & x = (E_P) \\
QR_m & x = (E_P \quad month)^T \\
QR_{LEDM} & x = (E_P \quad E_{P,M} \quad E_{P,V})^T \\
QR_{NAOI} & x = (E_P \quad NAOI)^T \\
QR_T & x = (T_P)
\end{array}
\begin{array}{l}
\left.\begin{array}{l} \\ \\ \\ \end{array}\right\} \to E_A \\
\} \to T_A
\end{array}
$$

7.2. Application to wind-assisted sailing propulsion

This section describes an example application for uncertainty estimation. To evaluate the quantile regression models the uncertainty of predictions for a Wind-assisted Ship Propulsion (WaSP) system is estimated. The propulsion energy is simulated for the route in fig. 6.2 (Europe → USA) with weather data for the time period from October 2005 to November 2010. Each prediction refers to a specific date d with a look-ahead time of h hours ($0 \leq h \leq 174$). Figure 7.1 shows the result of alg. 1. For this route $k=60$ is the optimal size of **roi**. It should be noted that the value of k depends on the specific route, the historical data that are available and the resolution of the weather data. However, fig. 7.1 shows that for this route any value of k greater than 30 improves the average normalized skill score. Figure 7.2 shows that there are basically no differences between the skill scores of QR_E, QR_m and QR_{NAOI}. The month and the NAOI do not improve the quality of the prediction intervals compared to QR_E. This shows that the season and the NAOI are too general information to be linked to prediction uncertainty in weather routing. The skill score for the QR_{LEDM} model is consistently higher for arbitrary values of α. Since the skill score takes into account both reliability (eq. 4.5) and sharpness (eq. 4.4) the *LEDM* clearly add relevant information to the model that is not provided by E_P (information about the route environment **roi**). Further benchmarks are shown in sec. 7.6.

7.2. Application to wind-assisted sailing propulsion

Figure 7.1: Normalized skill scores (eq. 4.6) of different significance levels α.

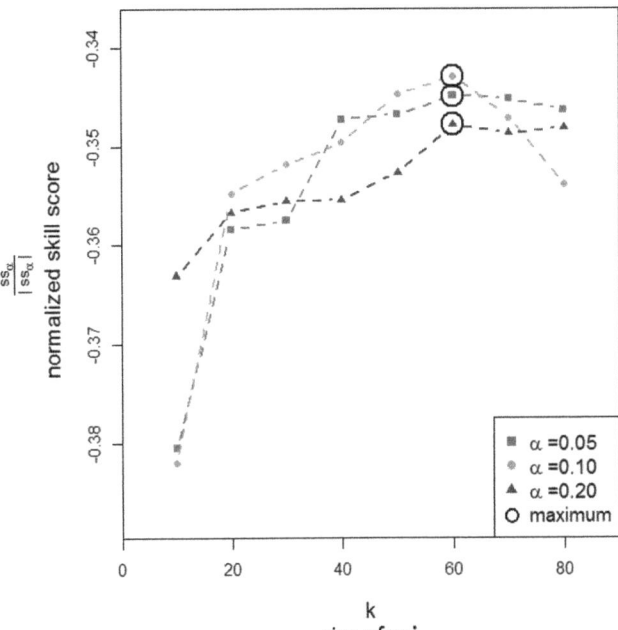

Figure 7.3 shows the skill scores of prediction intervals for the ship propulsion energy on the route in fig. 6.2 with different wind propulsion systems (WPS) and without a WPS. It is noted that the wind conditions are suitable for all three WPS on that route. It shows that a WPS significantly increases the uncertainty of predictions for the propulsion energy although this also depends on the parameterization of the WPS in appendix A. While the WPS reduces the ship propulsion energy by up to 40 % the prediction uncertainty of E_p (skill score) decreases by a factor of 2–3. The relatively high skill score of the DynaRig is due to the model (fig. 2.1) which assumes an optimal sail adjustment for individual wind conditions.

Figure 7.2: Scores for prediction intervals calculated with different QR models (tbl. 7.1). The prediction intervals are depicted in sec. 7.3.

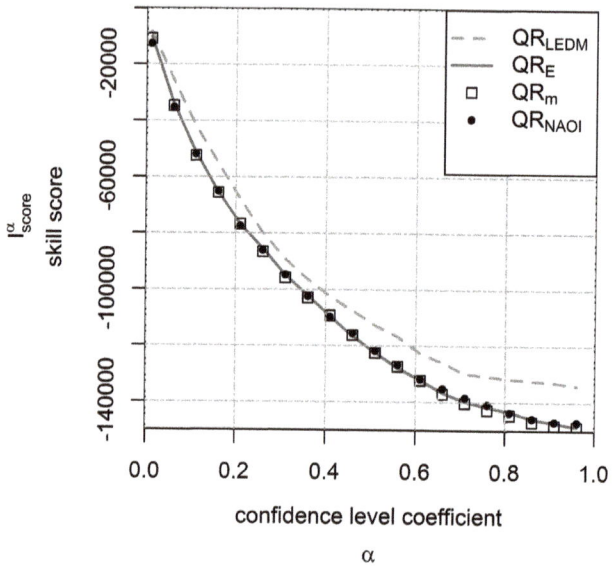

Figure 7.3: Averaged prediction uncertainty on a route from Wilhelmshaven to Baltimore

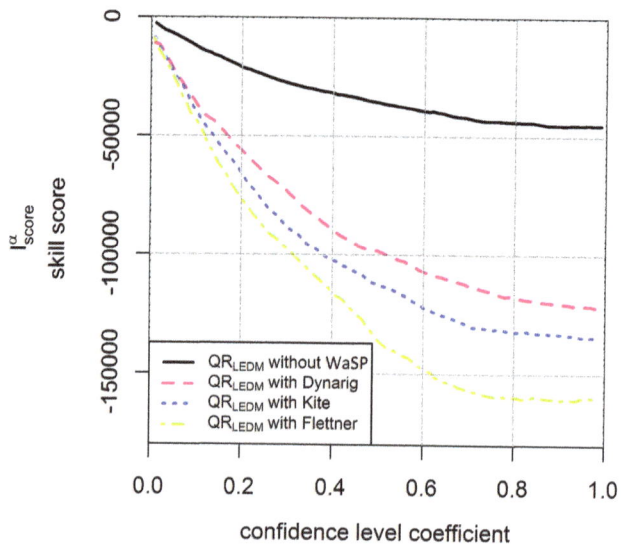

7.3. Prediction intervals for ship propulsion energy

Figures 7.4–7.9 show prediction intervals (I^α) for the ship propulsion energy (E_A) of a ship for a route from Wilhelmshaven to Baltimore for the years 2005 to 2010. The ship is the multi-purpose carrier described in sec. 2.2.2.1–2.2.5 (equipped with the Flettner-rotors described in sec. 2.2.1.3) and travels at a constant speed of 13 kn. The ship route is optimized with the A* algorithm described in sec. 6.2 (no uncertainty, strategy 1). E_A is calculated with weather analyses and predictions once per week.

The prediction intervals are calculated at the departure time of the ship. Each E_A value represents the propulsion energy of the ship for the complete route. It should be noted that the intervals are the main result of the work in this thesis. The diagrams on the left of each figure show prediction intervals which are calculated with the QR_E model and the diagrams on the right show prediction intervals which are calculated with the QR_{LEDM} model (see tbl. 7.1). The intervals which are calculated with QR_{LEDM} are significantly sharper while covering a similar percentage of E_A values compared with the prediction intervals calculated with the QR_E model. The intervals in fig. 7.4 (b) appear to be less smooth because these are calculated based on multiple parameters in contrast to the prediction intervals in fig. 7.4 (a).

It also shows that E_A is often much lower than E_P. This is because the route is re-optimized every 24 hours so that the route is often improved compared with the initially projected route at departure time. Another observation is that the intervals in fig. 7.4 (b) are centered above the E_A values whereas the intervals in fig. 7.4 (a) overestimate the E_A values. This can also be observed in fig. 7.5 (a) where the E_A value that corresponds to the highest E_P value is slightly greater than the lower boundary of I^α whereas the same E_A value in fig. 7.5 (b) is located close to the center of I^α.

Even though statistically E_A lies within I^α in both cases the distribution of E_A values within I^α can be important (depending on the application). Thus the QR_{LEDM} model reduces the bias of the prediction intervals. The use of LEDM lowers the probability of large prediction errors since the environment is taken into account, too. For example, the worst prediction error I^α_{Err} in fig. 7.6 (a) is 30 % lower in fig. 7.6 (b). Yet I^α_{Err} in fig. 7.6 (b) shows that prediction errors are not completely eliminated even with the QR_{LEDM} model.

In summary the QR_{LEDM} prediction intervals represent a more accurate estimate of the prediction uncertainty than the QR_E prediction intervals. This shows that the LEDM regressors represent significant information about prediction uncertainty in contrast to other variables like for example the NAO index, the time of the year or any other variable which has been evaluated in the course of this work.

Figure 7.4: 2005

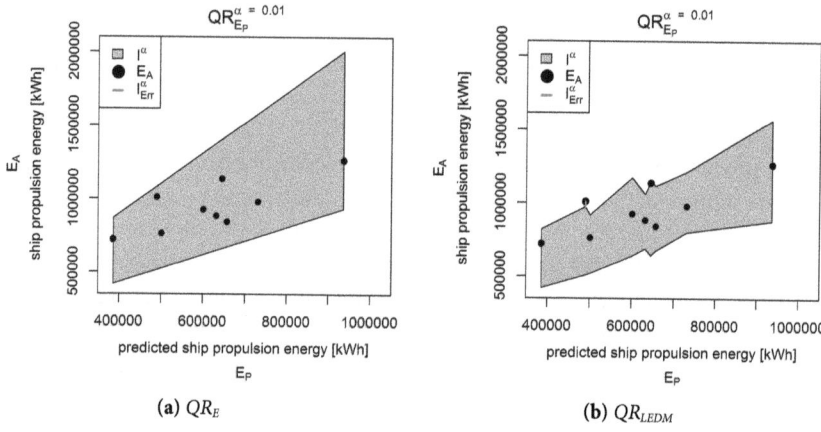

(a) QR_E (b) QR_{LEDM}

Figure 7.5: 2006

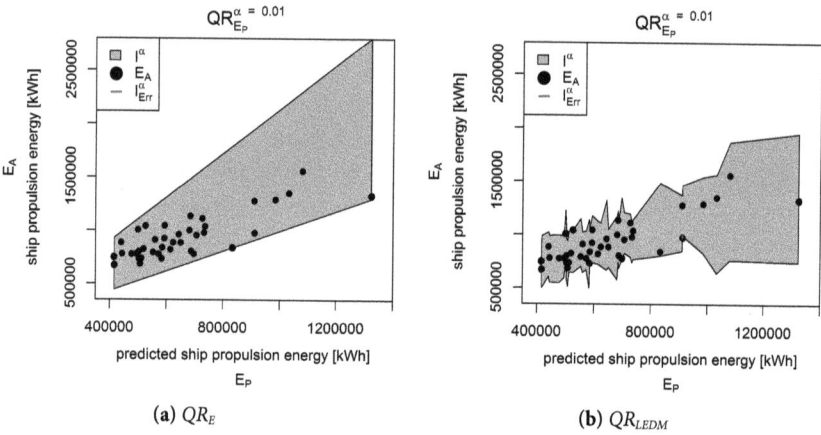

(a) QR_E (b) QR_{LEDM}

7.3. Prediction intervals for ship propulsion energy

Figure 7.6: 2007

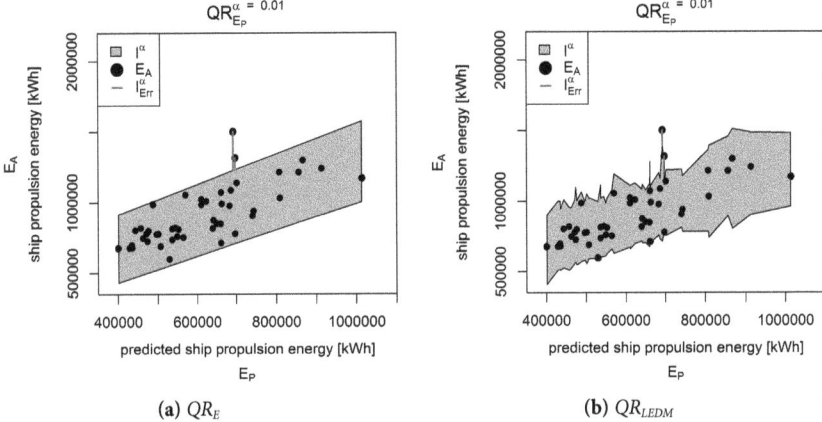

(a) QR_E (b) QR_{LEDM}

Figure 7.7: 2008

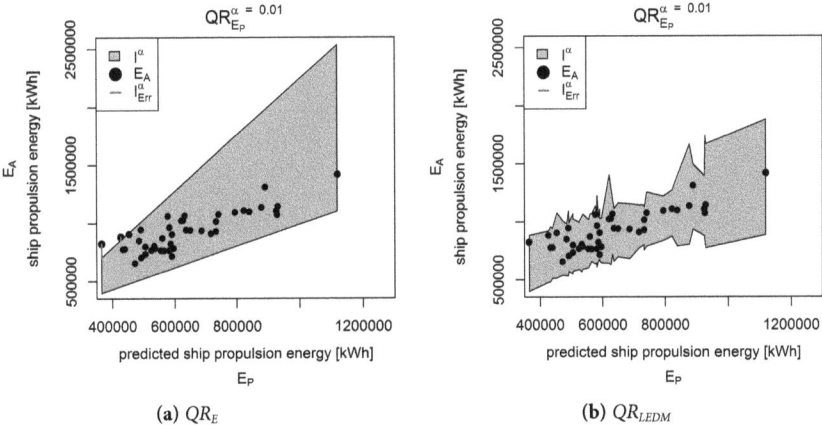

(a) QR_E (b) QR_{LEDM}

Figure 7.8: 2009

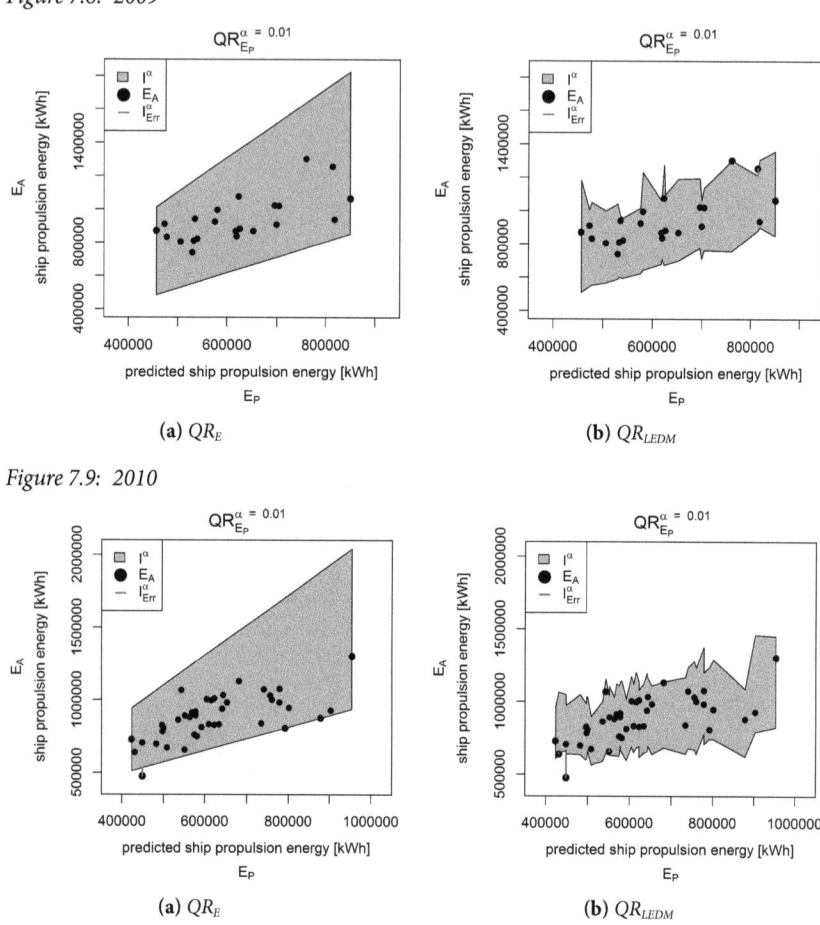

(a) QR_E (b) QR_{LEDM}

Figure 7.9: 2010

(a) QR_E (b) QR_{LEDM}

7.4. Prediction intervals for travel time

Figures 7.10–7.12 show prediction intervals I^α for the ship travel time T_A for a ship that travels a route from Wilhelmshaven to Baltimore for the years 2005 to 2010. The ship propulsion model and weather data are the same as above. Each interval I^α is calculated with the QR_T model in tbl. 3.4. The coverage rate of T_A values shows that the method is also applicable to estimate the uncertainty of predictions for the travel time thus showing the versatility of quantile regression. Here the actual travel time T_A is often less than the initially predicted travel time T_P because the route is re-optimized every 24 hours.

7.4. Prediction intervals for travel time

The travel time variability is low because the ship speed is set to a constant value of 13 kn in the simulation. The LEDM are not used here because they were designed to indicate uncertainty of energy predictions and not travel time predictions. However, the methodology to calculate LEDM regressors can be adopted to this case if speed curves are available, i.e. ship speeds for given wave heights and a fixed ship engine power. Then, a distribution of ship speeds with environmental weather could be calculated and Local Time Distribution moments (LTDM) could be calculated analogously to LEDM and be used as regressors in the QR model.

Figure 7.10: QR_T

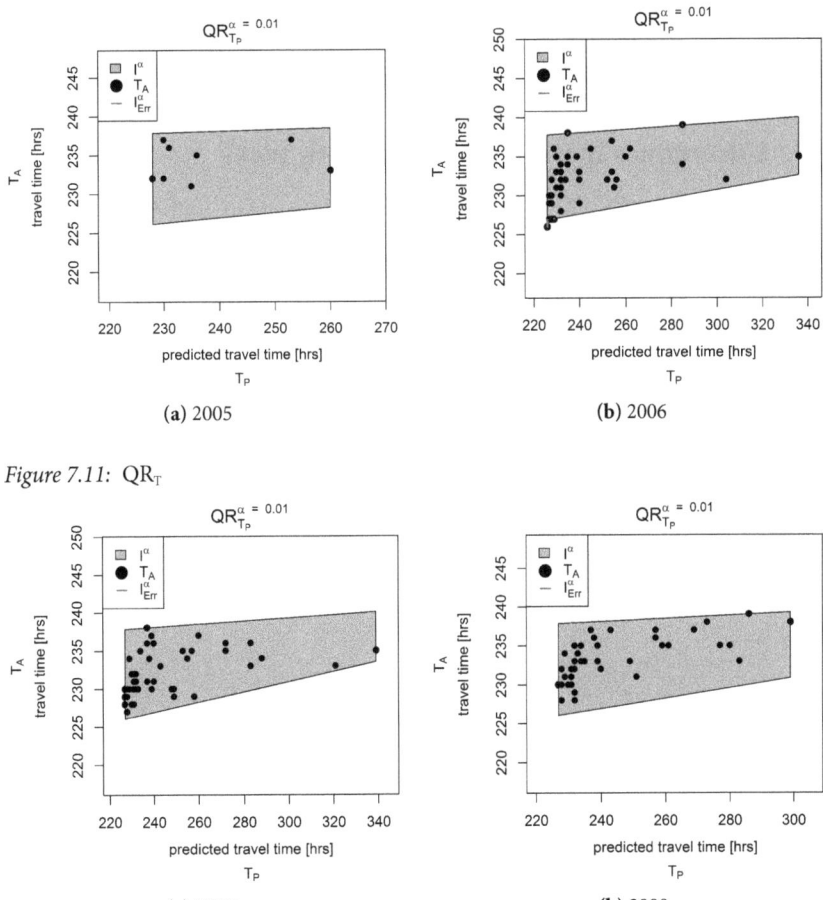

(a) 2005

(b) 2006

Figure 7.11: QR_T

(a) 2007

(b) 2008

Figure 7.12: QR_T

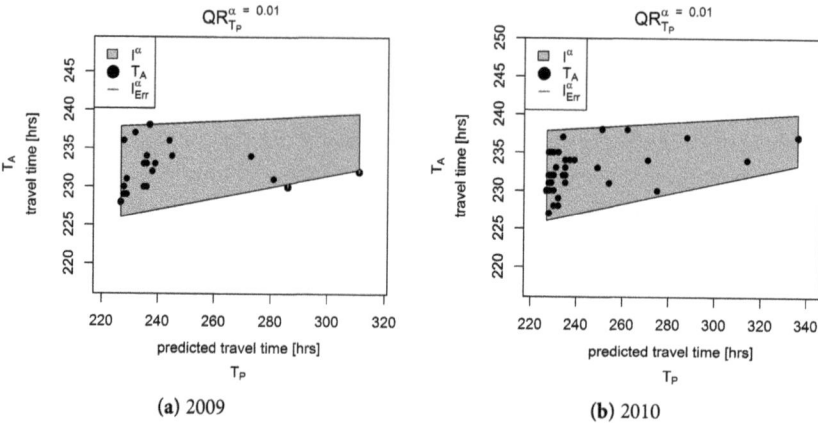

(a) 2009

(b) 2010

7.5. Uncertainty in weather routing with WaSP

The strategies described in sec. 6.3 yield different results as depicted in fig. 7.13. The upper histograms show that route optimization for a multi-purpose carrier with a wind propulsion system (WPS) can yield prediction errors which almost equal the mean ship propulsion energy for the complete route. Without a WPS the prediction error would be lower because the ship propulsion energy depends less on the weather conditions. In addition, the ship speed (13 kn) and the specific route influence the prediction error.

The histograms on the left show that taking into account prediction uncertainty (strategy 2 in alg. 4) reduces the average prediction error by 42 % (on a specific route with the available data). The average fuel consumption also decreases by 2 % because inefficient detouring happens less often. It should also be noted that the mean travel time decreases with strategy 2 (not depicted here) because unnecessary route optimization is avoided.

Figure 7.13: Statistics for prediction error Err_D (l.) and ship propulsion energy E_D (r.) (alg. 4) for route optimization with (b.) and without (u.) taking account prediction uncertainty (by using the shortest route R_1 if $I_U^\alpha > I_{U,Max}^\alpha$).

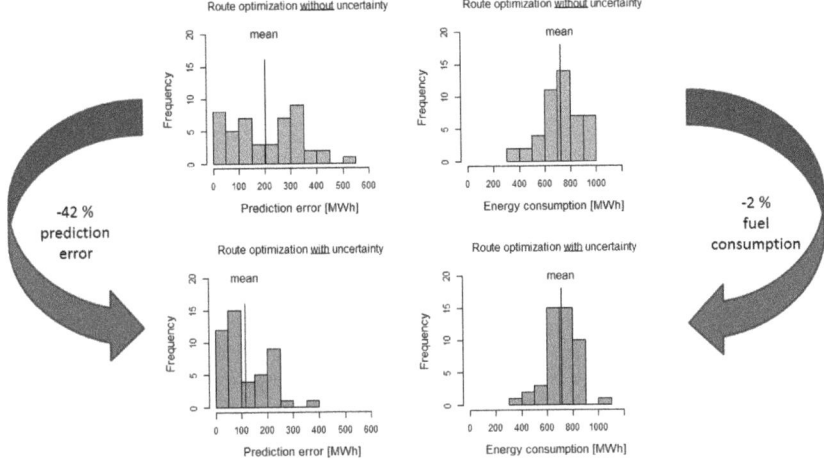

7.6. Prediction interval benchmarks

The visualization of prediction intervals is important to compare different methods. In order to confirm the results it is necessary to calculate statistical benchmarks. The skill score (eq. 4.6) is used here to assess the quality of prediction intervals since it assesses both reliability and sharpness of prediction intervals. The same data, i.e., weather data from the time period between 2005 and 2010, are used here to calculate prediction intervals and skill scores of different significance parameters α. A value of $\alpha = 0.01$ implies that the intervals are scaled to cover $100 - \alpha = 99$ % of the historical E_A values. As an alternative to quantile regression other non-parametric methods are available to calculate prediction intervals, e.g., least-squares support vector regression (LSSVR). The comparison between QR and LSSVR is of interest because the basic approaches of both methods are different: QR estimates a function (i.e., a conditional quantile) based on splines whereas LSSVR uses weighted kernel functions for function estimation. The LSSVR implementation from (Brabanter et al., 2011) with radial basis kernel functions is used here to calculate prediction intervals as a benchmark for the QR prediction intervals. The MATLAB code for LSSVR is freely available from Brabanter et al. (2013b).

Figure 7.14: Scores of prediction intervals calculated with quantile regression (QR) and least-squares support vector regression (LSSVR).

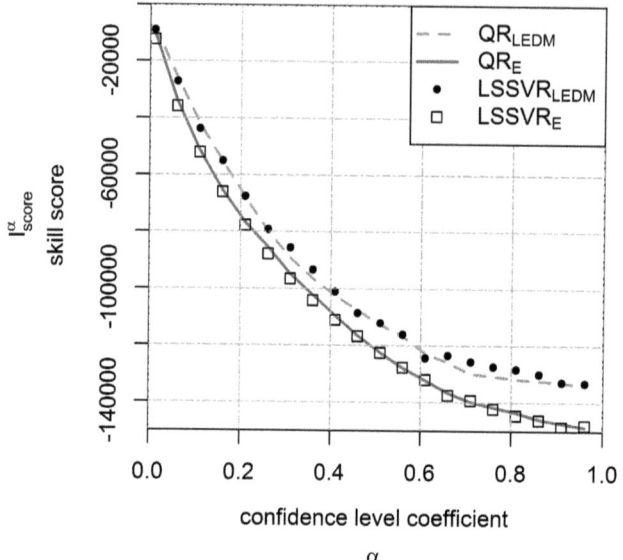

Figure 7.14 compares skill scores of prediction intervals calculated with QR and LSSVR. It shows that the skill scores are almost identical. Hence the LEDM improve the skill scores independently from the statistical method. Yet the runtime of the QR model ($\approx$ 1.2 *seconds*) is less than the runtime of the LSSVR model ($\approx$ 5 *seconds*) due to the data-driven parameter tuning and the estimation of the variance function in LSSVR. Due to this parameter tuning the LSSVR is less robust than the QR model: Depending on the data-driven tuning of the hyper-parameters (with cross-validation) the model does not always converge to a solution, i.e., does not calculate prediction intervals. Figure 7.15 confirms the visual impression of the prediction intervals in fig. 7.4–7.9. The RMSE shows that the percentage of large errors is significantly reduced with the QR_{LEDM} model. It also shows that the month (QR_m) and the NAOI (QR_{NAOI}) do not influence the average RMSE (eq. 4.3). This shows that this information does not indicate the level of prediction uncertainty when the predicted value E_p is known. In fact the current month is an indicator for prediction uncertainty (since it correlates with the number of storms), but E_p is a more specific information than the month or the NAOI. In contrast to that, the LEDM contain information about prediction uncertainty that is not provided by E_p, i.e., information about the energy variability in the local environment.

Figure 7.15: RMSE$_E$ *for different QR models. The average RMSE (eq. 3.4) of the* QR$_{LEDM}$ *model is up to 26 % lower compared with the other models.*

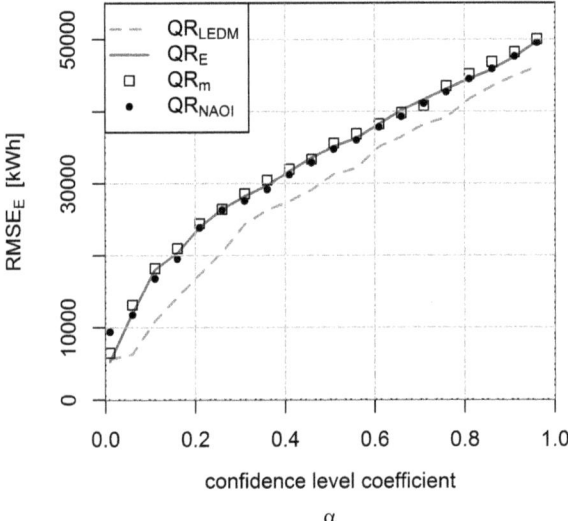

Figure 7.16: Skill scores of prediction intervals for the ship propulsion energy with different WPS technologies. The WPS drastically increases prediction uncertainty.

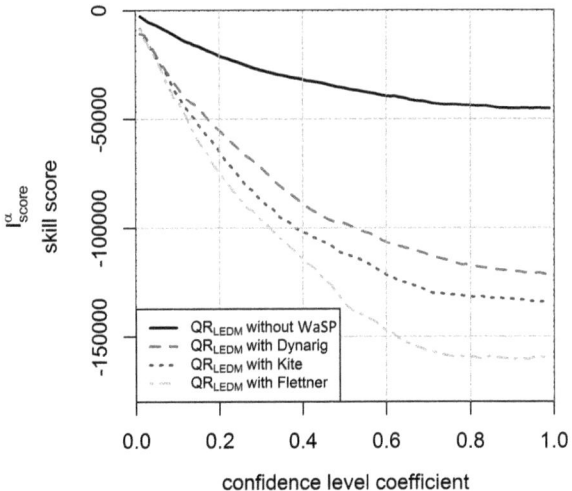

The motivation for this work was to investigate the effect of wind propulsion systems (WPS) on prediction uncertainty. Therefore, a method was developed to estimate prediction uncertainty accurately. Figure 7.16 shows that a WPS significantly increases prediction uncertainty. The findings imply that although a WPS can provide significant fuel savings the savings are difficult to predict if the predictions are based on long-term weather predictions.

7.7. Runtime discussion

Optimizing the quantile regression model (train_LSSVR in fig. C.5) and calculating the prediction interval for a route takes approximately one second on a standard PC (run_LSSVM in fig. C.5) which meets actual real-time requirements. The generation of the optimization data (simulate_historic_scenarios in fig. C.5) on the other hand can take several minutes with 5 years of historical data. The exact computation time depends on the efficiency of the database access and the calculation of the energy values. It is necessary to calculate a new set of optimization data if the predictions refer to a new site. Thus, for a wind turbine it suffices to calculate optimization data once. For weather routing a limited set of route waypoints can limit the number of possible routes. Then, the optimization data can be calculated in advance for each possible route.

7.8. Annual cost savings for a multi-purpose carrier

Figure 7.13 shows that taking into account prediction uncertainty reduces the average WaSP energy. The potential for savings depends on the current bunker price P_b, the ship speed V_s and the allowed travel time limit (see fig. 7.18) which is set in the weather routing application (see alg. 3). Equations 7.1a–7.1f calculate the annual cost savings for a multi-purpose-carrier, e.g., for the route in fig. 6.2 (distance $d_{route} = 3744$ sm).

$t_{year} = 365 \cdot 24$	[hrs]	hours per year
t_{route}	[hrs]	route travel time with speed V_s
P_{route,S_1}	[kWh]	mean annual propulsion energy with strategy S_1 (alg. 4)
P_{route,S_2}	[kWh]	mean annual propulsion energy with strategy S_2 (alg. 4)
P_Δ	[kWh]	mean annual propulsion energy savings
C_Δ	$[\frac{L}{h}]$	mean annual fuel consumption savings
M_Δ	[USD]	mean annual cost savings

7.8. Annual cost savings for a multi-purpose carrier

$$t_{route} = \frac{d_{route}}{V_s} \quad (7.1a)$$

$$P_{year,S_1} = P_{route,S_1} \cdot \frac{t_{year}}{t_{route}} \quad (7.1b)$$

$$P_{year,S_2} = P_{route,S_2} \cdot \frac{t_{year}}{t_{route}} \quad (7.1c)$$

$$P_\Delta = 1 - \frac{P_{route,S_2}}{P_{route,S_1}} \quad (7.1d)$$

$$C_\Delta = 0.185 \cdot P_{savings} \quad (7.1e)$$

$$M_\Delta = C_\Delta \cdot P_b \quad (7.1f)$$

Since the economical crisis in 2008 the bunker fuel prices varied significantly. Therefore, the economical consequence of the savings potential (2 %) in fig. 7.13 depends on the current bunker fuel price. Figure 7.17 shows the annual cost savings for historical bunker prices[12] during a period of ten years. Figure 7.18 shows the relation between the maximum allowed travel time and corresponding average WaSP energy. Since the ship speed V_s is set to a fixed value the maximum allowed travel time t_{max} is the degree of freedom that controls how far the ship detours off the great circle. The minimum travel time observed for 52 weekly simulation runs in 2008 is denoted by t_{min}, the maximum observed travel time is t_{max} respectively.

Figure 7.18 shows that the average ship propulsion power decreases by 17 % and the average travel time increases by up to approximately 2 days at the same time thus showing the potential of weather routing with WaSP.

---

12 Source for bunker prices: www.bunkerworld.com/, accessed 10.11.2015.

Figure 7.17: Annual fuel cost savings for historical bunker prices.

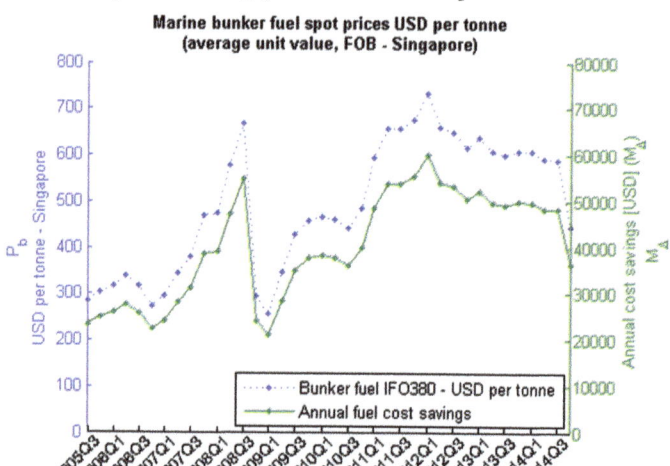

Figure 7.18: Maximum allowed travel time versus corresponding average ship propulsion energy on a route from Baltimore to Europe with a Flettner-rotor.

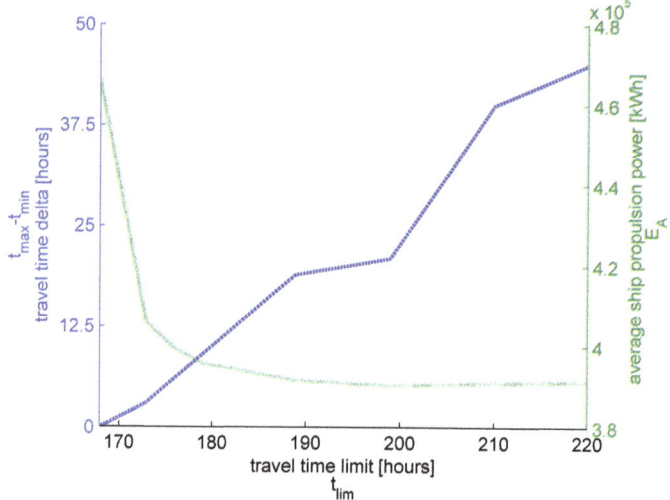

8. Conclusions and Outlook

Abstract: *This chapter concludes and discusses future developments within uncertainty estimation for weather-dependent energy predictions. The applicability of the quantile regression model to various other onshore and offshore applications is emphasized.*

Contents

8.1. New approach and scientific contribution 95
8.2. Outlook 96
    8.2.1. Prediction uncertainty of wind turbine power output 96
    8.2.2. Uncertainty in weather routing 96
    8.2.3. Further applications 97

8.1. New approach and scientific contribution

A quantile regression (QR) model was developed to estimate prediction intervals for weather-dependent energy predictions. The model is robust, meaning that it applies to arbitrary weather conditions, look-ahead times and sites. The QR model features two Local Energy Distribution Moments (LEDM) regressors which represent the mean and the variance of the energy calculated with weather predictions in the region of interest. The center of the region of interest can be for example

- a fixed local site, e.g., an offshore wind power plant or
- a waypoint along a shipping route.

This distinguishes the method from other approaches where the uncertainty is described by weather data. The method was integrated into an A* weather routing application. By skipping the route optimization (simulation of a route from the USA to Europe with one year of weather data) when only uncertain predictions were available the average statistical prediction error decreased by 42 % and the average ship propulsion energy also decreased by 2 %.

The LEDM regressors Weather-dependent energy predictions often depend on multiple weather parameters (wind speed, wind direction, etc.). Therefore, similar regressors from single weather parameters (e.g., mean local wind speeds) in (Pinson *et al.*, 2007a) did not improve the score of prediction intervals. The results with the LEDM regressors show that regressors for uncertainty estimation with quantile regression should be calculated based on the predicted energy and not based on individual weather parameters.

Weather routing is improved by uncertainty estimates It was shown that the results of weather routing can be improved if the route is only optimized (based on the predicted weather) if the weather prediction is reliable. The results might further improve if the uncertainty estimates are integrated into the route optimization procedure, e.g., by skipping partial routes for which only uncertain predictions exist. In addition, estimating the uncertainty of predictions for the propulsion energy and travel time enables better supply chain management and support the development of wind-assisted ship propulsion (WaSP) systems.

Prediction uncertainty with WaSP It was shown that a WPS increases prediction uncertainty drastically. The skill score for prediction intervals for the propulsion energy of the ship is up to 4 times lower for a WaSP model because inaccurate weather predictions have a greater impact on the ship's propulsion energy.

8.2. Outlook

8.2.1. Prediction uncertainty of wind turbine power output

Predictions for wind turbines and ship propulsion both depend on the weather prediction (differences are described in sec. 2.3). The LEDM regressors are variables that describe the uncertainty of the predicted energy in a region of interest. The LEDM can be used in a quantile regression model but also in other methods, e.g., as attributes in clustering (see. 4.3.1).

8.2.2. Uncertainty in weather routing

On-time delivery is an important requirement in (global) logistic supply chains. Yet predictions for ship travel time are less accurate than predictions for other vessel types (trucks, air planes) because of the uncertainty in weather predictions. With WaSP uncertainty in weather routing becomes even more important. The savings potential of WaSP might lead to greater deviations from the great circle which requires estimating the uncertainty of the prediction. The approach in this work can be used to predict other variables, e.g., the uncertainty of predictions for the travel time (as shown in the evaluation in sec. 7.4) or variables that describe ship stability.

8.2.3. Further applications

Prediction uncertainty is of interest in many applications. Non-linear dependencies between boundary conditions and the variables of interest (e.g., the relation between weather and ship propulsion energy) make it difficult to estimate prediction uncertainty (compare chap. 2). The approach in this work might be a blueprint for similar applications in this regard, e.g., growth predictions in economics.

Bibliography

Abramowitz, M., & Stegun, I. A. 1972. *Handbook of Mathematical Functions: with Formulas, Graphs, and Mathematical Tables.* 9th edn. Dover Publications: Mineola, NY, p. 79.

Ahrens, C. D., & Samson, P. 2010. *Extreme Weather and Climate.* Brooks Cole Publishing: Belmont, CA, p. 408.

Aschenbeck, S., Elsner, R., Folkerts, C., Ihnen, P., Lenger, T., Szczesn, W., Kreutzer, M., & Schlaak, M. 2009. *Forschung für den Klimaschutz und Schutz vor Klimawirkungen: Entwicklung eines integrierten Antriebs- und Nutzungskonzeptes auf Basis von Windkraft.* Project deliverable report 01LS05081. Bundesministerium für Bildung und Forschung: Jade Hochschule Oldenburg, pp. 14–15.

Beeken, A. 2010. FINO1-platform: Operation and Data Analysis of an Offshore based LIDAR Device. *DEWI Magazine,* **36**(8), pp. 53–62.

Bentin, M., Zastrau, D., Schlaak, M., Freye, D., Elsner, R., & Kotzur, S. 2016. A New Routing Optimization Tool-influence of Wind and Waves on Fuel Consumption of Ships with and without Wind Assisted Ship Propulsion Systems. *Ocean Engineering,* **14**(1), pp. 153–162.

Bessa, R. J., Miranda, V., Botterud, A., Wang, J., & Constantinescu, E. M. 2012. Time Adaptive Conditional Kernel Density Estimation for Wind Power Forecasting. *IEEE Transactions on Sustainable Energy,* **3**(4), pp. 660–669.

Blendermann, W. 1996. *Wind Loading of Ships: Collected Data from Wind Tunnel Tests in Uniform Flow.* Technical report RA489(574). Institut für Schiffbau der Universität Hamburg: TUHH Universitätsbibliothek, pp. 1–62.

Blume, P. 1977. *Berechnung des Seeverhaltens für eine systematisch variierte Formfamilie.* Technical report 1505. Hamburgische Schiffbau-Versuchsanstalt: TUHH Universitätsbibliothek, pp. 1–35.

Borge, N. J. C., Hessner, K., & Reichert, K. 1999. Estimation of the Significant Wave Height with X-Band Nautical Radars (contribution OMAE99/OSU-3063). In: Chakrabarti, S. K. (ed), *Proceedings of the 18th International Conference on Offshore Mechanics and Arctic Engineering (OMAE),* July 11–16, 1999, St John's, NF. ASME: New York, NY, pp. 1–8.

Böttcher, J. 2013. *Handbuch Offshore-Windenergie: Rechtliche, technische und wirtschaftliche Aspekte.* De Gruyter Oldenbourg: München, p. 450.

Bouckaert, R., Frank, E., Holmes, G., & Fletcher, D. 2011. A comparison of methods for estimating prediction intervals in NIR spectroscopy: Size matters. *Chemometrics and Intelligent Laboratory Systems,* **109**(2), pp. 139–145.

Bowditch, N. 2011. *American Practical Navigator: An Epitome of Navigation and Nautical Astronomy*. National Imagery and Mapping Agency: Bethesda, MD, pp. 542–547.

Brabanter, K., Brabanter, J., Suykens, J. A., & Moor, B. 2011. Approximate confidence and prediction intervals for least squares support vector regression. *IEEE transactions on neural networks*, **22**(1), pp. 110–120.

Brabanter, K., Suykens, J., & Moor, B. 2013a. Nonparametric regression via StatLSSVM. *Journal of Statistical Software*, **55**(2), pp. 1–22.

Brabanter, K. De, Suykens, J. A. K., & De Moor, B. 2013b. Nonparametric regression via StatLSSVM. *Journal of Statistical Software*, **55**(2), pp. 1–21.

Bremnes, J. B. 2004. Probabilistic wind power forecasts using local quantile regression. *Wind Energy*, **7**(1), pp. 47–54.

Chan, T. F., Golub, G. H., & Leveque, R. J. 1983. Algorithms for Computing the Sample Variance: Analysis and Recommendations. *The American Statistician*, **37**(3), p. 242.

Charnock, H. 1955. Wind stress on a water surface. *Quarterly Journal of the Royal Meteorological Society*, **81**(350), pp. 639–640.

Chatfield, C. 1993. Calculating Interval Forecasts. *Journal of Business & Economic Statistics*, **11**(2), pp. 121–135.

Chatfield, C. 2001. *Prediction intervals for time series*. Kluwer Academic Publishers: Norwell, MA, pp. 475–494.

Christoffersen, P. F. 1998. Evaluating Interval Forecasts. *International Economic Review*, **39**(4), pp. 841–862.

Dempster, A. P., Laird, N. M., & Rubin, D. B. 1977. Maximum Likelihood from Incomplete Data via the EM Algorithm. *Journal of the Royal Statistical Society: Series B (Statistical Methodology)*, **39**(1), pp. 1–38.

Ehrendorfer, M. 1997. Predicting the uncertainty in numerical weather forecasts: A review. *Meteorologische Zeitschrift*, **6**(4), pp. 147–183.

France, W., Levadou, M., Treakle, T. W., Paulling, J. R., Michel, R. K., & Moore, C. 2003. An investigation of head-sea parametric rolling and its influence on container lashing systems. *Marine Technology*, **40**(1), pp. 1–19.

Gneiting, T., & Raftery, A. E. 2007. Strictly Proper Scoring Rules, Prediction, and Estimation. *Journal of the American Statistical Association*, **102**(477), pp. 359–378.

Görges, M., Möller, J., & Shao, J. 2014. Simulationsgestützte Planung und Steuerung in der Offshore-Logistik. In: Thoben, K. D., Haasis, H.-D., & Lewandowski, M. (eds), *Logistik für die Windenergie: Herausforderungen und Lösungen für moderne Windkraftwerke: Industrie-Symposium*, December 3, 2014, Bremen. ePubli: Berlin, pp. 49–58.

Greengard, S. 2014. Weathering a New Era of Big Data. *Communications of the ACM*, **57**(9), pp. 12–14.

Greiner, S., & Joschko, P. 2014. Analyse von Instandhaltungsprozessen im Leistungssystem Offshore-Windpark. In: Thoben, K. D., Haasis, H.-D., & Lewandowski, M. (eds), *Logistik für die Windenergie: Herausforderungen und Lösungen für moderne Windkraftwerke: Industrie-Symposium*, December 3, 2014, Bremen. ePubli: Berlin, pp. 87–98.

Hagedorn, R., Doblas-Reyes, F. J., & Palmer, T. N. 2005. The rationale behind the success of multi-model ensembles in seasonal forecasting. *Tellus A*, **57**(3), pp. 219–233.

Hagiwara, H., & Spaans, J. A. 1987. Practical Weather Routing of Sail-assisted Motor Vessels. *Journal of Navigation*, **40**(1), pp. 96–119.

Härdle, W. 2004. *Nonparametric and Semiparametric Models*. Springer: Berlin, pp. 241–243.

Hau, E. 2008. *Windkraftanlagen: Grundlagen, Technik, Einsatz, Wirtschaftlichkeit*. 4th edn. Springer: Berlin, pp. 515–517.

Heidmann, R. 2014. Losgröße 1plus – Logistikmanagement im schweren Maschinen- und Anlagenbau mit geringen Losgrößen und oftmals prototypischen Rahmenbedingungen. In: Thoben, K. D., Haasis, H.-D., & Lewandowski, M. (eds), *Logistik für die Windenergie: Herausforderungen und Lösungen für moderne Windkraftwerke: Industrie-Symposium*, December 3, 2014, Bremen. ePubli: Berlin, pp. 25–34.

Hellmann, G. 1914. *Über die Bewegung der Luft in den untersten Schichten der Atmosphäre*. Sitzungsbericht Band 1. Reimer Verlag: Staatsbibliothek zu Berlin, pp. 415–437.

Hoare, C. A. R. 1961. Algorithm 65: Find. *Communications of the ACM*, **4**(7), pp. 321–322.

Hoffschildt, M., Bidlot, J.-R., Hansen, B., & Janssen, P. 1999. Potential benefit of ensemble forecasts for ship routing. *Technical Memorandum*, **1**(287), pp. 1–25.

Hoschek, J. 1984. *Mathematische Grundlagen der Kartographie*. 2nd edn. Mannheim: Bibliographisches Institut, pp. 180–194.

Hurrell, J. W., Kushnir, Y., Ottersen, G., & Visbeck, M. 2003. *The North Atlantic oscillation: climatic significance and environmental impact*, Eds. Geophysical Monograph Series, *134*, 279. Washington, DC: American Geophysical Union.

James, R. W. 1957. *Application of wave forecast to marine navigation*. Special Publication SP-1. US Navy Hydrographic Office: Washington, D.C., pp. 1–85.

Jarabo-Amores, P., La Mata-Moya, D. de, Hessner, K., & Nieto-Borge, J. C. 2008. Signal-to-noise ratio analysis to estimate ocean wave heights from X-band marine radar image time series. *IET Radar, Sonar & Navigation*, **2**(1), pp. 35–41.

Jehan, T. 2005. *Creating Music by Listening*. Thesis (PhD), Massachusetts Institute of Technology (MIT), p. 83.

Jolliffe, I. T. 2002. *Principal Component Analysis*. 2nd edn. Springer: New York, NY, pp. 1–488.

Jones, M. T. 2015. *Artificial Intelligence: A Systems Approach*. Infinity Science: Hingham, MA, pp. 57–58.

Jovanoski, Z., & Robinson, G. 2009. Ship stability and parametric rolling. *Australasian Journal of Engineering Education*, **15**(2), p. 43.

Juban, J., Siebert, N., & Kariniotakis, G. 2007. Probabilistic Short-term Wind Power Forecasting for the Optimal Management of Wind Generation. *In:* Bialek, J. (ed), *Proceedings of the IEEE Power Tech Conference*, July 1–5, 2007, Lausanne. IEEE: Newry, pp. 683–688.

Khan, M. Z. A. 1998. *Text Book Of Practical Geography*. Concept Publishing: New Delhi, p. 62.

Kleemann, M., & Meliß, M. 1993. *Regenerative Energiequellen: Mit 75 Tabellen*. 2nd edn. Springer: Berlin, pp. 959–988.

Klessmann, C., Held, A., Rathmann, M., & Ragwitz, M. 2011. Status and perspectives of renewable energy policy and deployment in the European Union-What is needed to reach the 2020 targets? *Energy policy*, **39**(12), pp. 7637–7657.

Koenker, R. W. 2015. *quantreg: Quantile Regression*. R package version 5.11. Available from http://CRAN.R-project.org/package=quantreg (accessed 14.02.2015).

Koenker, R. W., & Bassett, G. 1978. Regression Quantiles. *Econometrica*, **46**(1), pp. 33–50.

Kunz, H., Hagens, N., & Balogh, S. 2014. The Influence of Output Variability from Renewable Electricity Generation on Net Energy Calculations. *Energies*, **7**(1), p. 161.

Lange, B., Larsen, S., Højstrup, J., & Barthelmie, R. 2004. Importance of thermal effects and sea surface roughness for offshore wind resource assessment. *Journal of Wind Engineering and Industrial Aerodynamics*, **92**(11), pp. 959–988.

Li, Y., & Zhu, J. 2008. L1-Norm Quantile Regression. *Journal of Computational and Graphical Statistics*, **17**(1), pp. 163–185.

Lorenz, E. N. 1969. A study of the predictability of 28-variable atmosphere model. *Tellus A*, **21**(21), pp. 739–759.

Magnus, G. 1853. Über die Abweichung der Geschosse, und: Ueber eine auffallende Erscheinung bei rotirenden Körpern. *Annalen der Physik und Chemie*, **164**(1), pp. 1–29.

Mahmud, M. 2004. Skill of a superensemble forecast over equatorial Southeast Asia. *International Journal of Climatology*, **24**(15), pp. 1963–1972.

Maki, A., Akimoto, Y., Nagata, Y., Kobayashi, S., Kobayashi, E., Shiotani, S., Ohsawa, T., & Umeda, N. 2011. A new weather-routing system that accounts for ship stability based on a real-coded genetic algorithm. *Journal of Marine Science and Technology*, **16**(3), pp. 311–322.

MAN. 2011. *Basic Principles of Ship Propulsion*. White paper 5510–0004–02ppr. Available from marine.man.eu/docs/librariesprovider6/propeller-aftship/basic-principles-of-propulsion.pdf (accessed 03.09.2015).

Nielsen, H. A., Madsen, H., & Nielsen, T. S. 2006. Using quantile regression to extend an existing wind power forecasting system with probabilistic forecasts. *Wind Energy*, **9**(1–2), pp. 95–108.

Ouchi, K., Uzawa, K., & Kana, A. 2011. Huge Hard Wing Sails for the Propulsor of Next Generation Sailing Vessel. In: Abdel-Maksoud, M. (ed), *Second International Symposium on Marine Propulsors (SMP11)*, June 15–17, 2011, Hamburg. Buchwerft Verlag: Kiel, pp. 546–552.

Palmer, T. N., Shutts, G. J., Hagedorn, R., Doblas-Reyes, F. J., Jung, T., & Leutbecher, M. 2005. Representing model uncertainty in weather and climate prediction. *Annual Review of Earth and Planetary Sciences*, **33**(1), pp. 163–193.

Pinson, P. 2012. Very-short-term probabilistic forecasting of wind power with generalized logit-normal distributions. *Journal of the Royal Statistical Society: Series C (Applied Statistics)*, **61**(4), pp. 555–576.

Pinson, P., & Kariniotakis, G. 2004. On-line assessment of prediction risk for wind power production forecasts. *Wind Energy*, **7**(2), pp. 119–132.

Pinson, P., Nielsen, H. A., Madsen, H., Lange, M., & Kariniotakis, G. 2007a. *Methods for the estimation of the uncertainty of wind power forecasts*. Project deliverable report D3.1b. ANEMOS project: Technical University of Denmark, p. 56.

Pinson, P., Nielsen, H. A., Møller, J. K., Madsen, H., & Kariniotakis, G. N. 2007b. Non-parametric probabilistic forecasts of wind power: required properties and evaluation. *Wind Energy*, **10**(6), pp. 497–516.

Prölls, W. 1970. *Wind loading on Improvements in or relating to sailing vessels*. Patent specification DE1089656B. Intellectual Property Office: London.

Rinne, A., & Haasis, H.-D. 2014. Der Einsatz von Helikoptern im Service von Offshore-Windenergieanlagen. In: Thoben, K. D., Haasis, H.-D., & Lewandowski, M. (eds), *Logistik für die Windenergie: Herausforderungen und Lösungen für moderne Windkraftwerke: Industrie-Symposium*, December 3, 2014, Bremen. ePubli: Berlin, pp. 107–114.

Rogers, J. C. 1997. North Atlantic storm track variability and its association to the North Atlantic Oscillation and climate variability of northern Europe. *Journal of Climate*, **10**(7), pp. 1635–1647.

Schlaak, M., Kreutzer, R., & Elsner, R. 2009. Simulating Possible Savings of the Skysails-System on International Merchant Ship Fleets. *The International Journal of Maritime Engineering*, **151**(A4), pp. 1–25.

Schlalos, I., Harenslak, T., Oelker, S., & Lewandowski, M. 2014. Technologien für die automatisierte Logistik in der Betriebsphase von Offshore-Windenergieanlagen. *In:* Thoben, K. D., Haasis, H.-D., & Lewandowski, M. (eds), *Logistik für die Windenergie: Herausforderungen und Lösungen für moderne Windkraftwerke: Industrie-Symposium*, December 3, 2014, Bremen. ePubli: Berlin, pp. 99–106.

Scholz-Reiter, B., Dashkovskiy, S., Görges, M., & Naujok, L. 2011. Stability analysis of autonomously controlled production networks. *International Journal of Production Research*, **49**(16), pp. 4857–4877.

Schütt, H., & Lange, K. 2014. Ein Logistik-Diagnose-Werkzeug für die Offshore-Windenergie. *In:* Thoben, K. D., Haasis, H.-D., & Lewandowski, M. (eds), *Logistik für die Windenergie: Herausforderungen und Lösungen für moderne Windkraftwerke: Industrie-Symposium*, December 3, 2014, Bremen. ePubli: Berlin, pp. 69–74.

Scott, D. B., Collins, E. S., Gayes, P. T., & Wright, E. 2003. Records of prehistoric hurricanes on the South Carolina coast based on micropaleontological and sedimentological evidence, with comparison to other Atlantic Coast records. *Geological Society of America Bulletin*, **115**(9), pp. 1027–1039.

Shao, W., Zhou, P., & Thong, S. K. 2012. Development of a novel forward dynamic programming method for weather routing. *Journal of Marine Science and Technology*, **17**(2), pp. 239–251.

Silverman, B. W. 1986. *Density Estimation for Statistics and Data Analysis*. Chapman and Hall: London, p. 48.

Spaans, J. A., & Stoter, P. H. 1995. New developments in ship weather routing. *Journal of Navigation*, **43**(169), pp. 95–106.

Stensrud, D. J., Brooks, H. E., Du, J., Tracton, M. S., & Rogers, E. 1999. Using Ensembles for Short-Range Forecasting. *Monthly Weather Review*, **127**(4), pp. 433–446.

Stull, R. B. 1988. *An Introduction to Boundary Layer Meteorology*. Book Publishing: Dordrecht, pp. 378–379.

Tambke, J., Claveri, L., Bye, J. A. T., Poppinga, C., Lange, B., Bremen, L. von, Durante, F., & Wolff, J. O. 2006. Offshore Meteorology for Multi-Mega-Watt Turbines. *In:* Beurskens, J., & Snel, H. (eds), *Proceedings of the European Wind*

Energy Conference (EWEC), February 27-March 2, 2006, Athens. European Wind Energy Association (EWEA), Red Hook, NY, pp. 139-143.

Traumüller, A. 2014. *Wellenwiderstand auf das Schiff.* Thesis (Bachelor), University of Applied Sciences Emden-Leer, p. 1-76.

Tsujimoto, M., & Tanizawa, K. 2006. Development of a Weather Adaptive Navigation System Considering Ship Performance in Actual Seas. *In:* Kuehnlein, W.L. (ed), *Proceedings of the 25th International Conference on Offshore Mechanics and Arctic Engineering (OMAE2006)*, June 4-9, 2006, Hamburg. ASME: Washington, D.C., pp. 413-421.

Wagner, B. 1966. *Windkanalversuche mit gewölbten Plattensegeln mit Einzelmasten sowie mit Plattensegel bei Mehrmastanordnung*. Monograph 171B. Institut für Schiffbau der Universität Hamburg: TUHH Universitätsbibliothek, pp. 1-85.

Wagner, C., Andersson, G., A., Raulien, I., Sauer, & M., Bellon. 1985. *Weiterentwicklung des Flettner-Rotors zum modernen Windzusatzantrieb*. Project deliverable report MTK 03084. Bundesministerium für Bildung und Forschung: Berlin, pp. 1-28.

Walter, K., & Graf, H.-F. 2005. The North Atlantic variability structure, storm tracks, and precipitation depending on the polar vortex strength. *Atmospheric Chemistry and Physics*, 5(1), pp. 239-248.

Wang, X., Guo, P., & Huang, X. 2011. A review of wind power forecasting models. *Energy procedia*, 12(1), pp. 770-778.

Zarnani, A., & Musilek, P. 2013. Learning uncertainty models from weather forecast performance databases using quantile regression. *In:* Szalay, A., Budavari, T., Balazinska, M., Meliou, A., & Sacan, A. (eds), *Proceedings of the 25th International Conference on Scientific and Statistical Database Management (SSDBM)*, July 29-31, 2013, Baltimore, MD. ACM: New York, NY, pp. 117-125.

Zarnani, A., Musilek, P., & Heckenbergerova, J. 2014. Clustering numerical weather forecasts to obtain statistical prediction intervals. *Meteorological Applications*, 21(3), pp. 605-618.

Zhi, X., Qi, H., Bai, Y., & Lin, C. 2012. A comparison of three kinds of multimodel ensemble forecast techniques based on the TIGGE data. *Acta Meteorologica Sinica*, 26(1), pp. 41-51.

List of Figures

1.1	Aggregate electricity demand in Denmark	1
2.1	Normalized propulsion force of the DynaRig for different wind angles	11
2.2	The Maltese Falcon super yacht	12
2.3	Normalized propulsion force of a kite-sail for different wind angles	13
2.4	The E-Ship 1	14
2.5	The normalized propulsion force of the Flettner-rotor for different wind angles	15
2.6	Calculation of the ship energy consumption for a ship with an optional wind propulsion system	19
2.7	Annual average official track errors of Atlantic basin tropical storms and hurricanes for the period 1970–2012	23
2.8	Comparison between predicted and true wind speeds	26
3.1	Partitioning of the North Atlantic Ocean	31
3.2	Average monthly significant wave height prediction error in the North Atlantic Ocean	33
3.3	The FINO research platforms in the North and Baltic Seas	35
3.4	Schema of the WaMoS II system	38
3.5	Average wave prediction accuracy at FINO 1 in the Northern Sea	39
3.6	Average wind prediction accuracy at FINO 1 in the Northern Sea	40
3.7	Average monthly wave prediction accuracy at FINO 1 in the Northern Sea	41
3.8	Wind speed prediction accuracy at FINO 1 in April 2010	42
3.9	GME wind speed bias at FINO 1	43
3.10	A comparison between the datum time prediction and one-week predictions	44
3.11	Deviation between DWD wind speed data and FINO 1 measurements	45
3.12	Average prediction errors between 2005 and 2010 for wind speed at FINO 1–3	46
3.13	Average prediction errors at FINO 1 for individual months	47

List of Figures

5.1	Schematic diagram of conditional quantiles	58
5.2	The check function for quantile regression	59
5.3	Principal component analysis after two iterations	61
5.4	The North Atlantic Oscillation pattern	62
5.5	The region of interest with respect to a single waypoint	67
6.1	The A* search algorithm after 2 iteration steps	74
6.2	The region of interest with respect to a complete shipping route	75
7.1	Normalized skill scores of different significance levels	81
7.2	Scores for prediction intervals calculated with different QR models	82
7.3	Averaged prediction uncertainty on a route from Wilhelmshaven to Baltimore	82
7.4	Quantile regression with data from 2005	84
7.5	Quantile regression with data from 2006	84
7.6	Quantile regression with data from 2007	85
7.7	Quantile regression with data from 2008	85
7.8	Quantile regression with data from 2009	86
7.9	Quantile regression with data from 2010	86
7.10	Quantile regression for travel time uncertainty with data from 2005–2006	87
7.11	Quantile regression for travel time uncertainty with data from 2007–2008	87
7.12	Quantile regression for travel time uncertainty with data from 2009–2010	88
7.13	Statistics for the prediction error and ship propulsion energy	89
7.14	A comparison between quantile regression (QR) and least-squares support vector regression (LSSVR)	90
7.15	Root-mean square errors for different QR models	91
7.16	Skill scores of prediction intervals for the ship propulsion energy with different WPS technologies	91
7.17	Annual fuel cost savings for historical bunker prices	94
7.18	Maximum allowed travel time versus corresponding average ship propulsion energy	94

- C.1 UML diagramm: the route class 117
- C.2 UML diagramm: the world model classes 118
- C.3 UML diagramm: the ship model class 119
- C.4 UML diagramm: the forecast heuristics (class diagram) 120
- C.5 UML diagramm: the forecast heuristics (sequence diagram) 121
- C.6 UML diagramm: the data model classes 122

List of Tables

2.1	Efficiency coefficients of the BBC Hudson	16
2.2	Common categories of weather forecast horizons	21
3.1	DWD weather forecast models	32
3.2	Positions of FINO platforms	35
3.3	FINO instruments and measurement heights	35
3.4	Indicators for prediction uncertainty	48
5.1	Estimators for statistical distribution moments	65
5.2	Quantile regression models for energy prediction	69
7.1	Quantile regression models for energy and time prediction	80
A.1	Parameters of the speed power curve model	113
A.2	Parameters of the Flettner-rotor model	113
A.3	Parameters of the Dyna-ship model	114
A.4	Parameters of the kite model	114
B.1	Wind resistance model parameter values of the BBC Hudson	115
B.2	Parameters of the wave resistance model	115
D.1	Outputs of the GSM model	123
D.2	Update intervals of the GSM model	123

A. Parameters of Wind Propulsion Systems

In order to achieve a realistic simulation of the ship propulsion energy the parameter values in this section are similar to the dimensions of wind propulsion systems and of a multi-purpose carrier which have actually been constructed and operated in reality. The parameter values of the speed power curve (sec. 2.2.2) in tbl. A.1 were fitted based on vessel speed and fuel consumption measurements conducted on the "BBC Hudson" (a multi-purpose carrier).

Table A.1: Parameters of the speed power curve model

description	symbol	units	value
slope in linear model	a_1	none	0.021
offset in linear model	a_2	none	0.286
ship draft	T	[m]	7
ship speed	V_s	[kn]	13

The values in tbl. A.2 are similar to the dimensions of the Flettner-rotors installed on the E-Ship 1 in fig. 2.4. The values in tbl. A.3 are similar to the dimensions of the DynaRig installation on the "Maltese Falcon" in fig. 2.2. And the values in tbl. A.4 are similar to the dimensions of the kite-sail has been installed on the "BBC SkySails" Aschenbeck et al. (2009).

Table A.2: Parameters of the Flettner-rotor model

name	symbol	units	value
number of rotors	N_R	none	4
height of rotor	H	m	25
diameter of rotor	D	m	4
area of rotor	A	m²	$A = H \cdot D$
diameter of disc on top of rotor	DS	m	8
average height of windage area	h	m	30
maximum rotation speed	$n_{rot,max}$	min^{-1}	200
maximum force on rotor	F_{max}	N	100000
maximum revolutions per minute	U_{max}	$\frac{r}{min}$	$\frac{\pi}{60} \cdot D \cdot n_{max}$
propulsion coefficient	ANFL	none	2.5
efficiency factor of electric drive	WKG	none	0.9

Table A.3: Parameters of the Dyna-ship model (with 3 DynaRigs)

name	symbol	units	value
sail area	A	m^2	800
average height of windage area	h	m	30
number of masts	N_M	integer	3

Table A.4: Parameters of the kite model

name	symbol	units	value
area of kite sail	A	m^2	800
average working height of kite	h	m	150
power usage of kite	PKI	W	2000
efficiency factor of electric drive	WKG	none	0.9

B. Parameters of the wind and wave resistance models

The parameters of the wind and wave models are similar to the dimensions of the "BBC Hudson" (a multi-purpose carrier). The wind and wave values for the simulation are historical weather data by the German Weather Service.

Table B.1: Wind resistance model parameter values of the BBC Hudson

name	symbol	units	value
air density	ρ_{air}	$\frac{kg}{m^3}$	1.25
cross factor	δ	N	0.4
lateral windage area	A_L	m^2	1690
frontal windage area	A_F	m^2	530
air resistance coeff. for lateral wind	CD_t	none	0.85
air resistance coeff. for longitudinal wind	CD_{lAF}	none	0.6
width of ship	B	m	22.8

Table B.2: Parameters of the wave resistance model

name	symbol	units	value
air density	ρ_{air}	$\frac{kg}{m^3}$	0.125
ship speed	V_s	kn	13
ship length	L	m	133
ship width	B	m^2	22.8
ship course	S_c	°	variable
wave direction	β	°	variable
rel. wave direction	γ	°	$180 - |(S_c - \beta)|$

C. UML diagrams of the Java implementation

Figure C.1: *The class* `Route` *encapsulates the ship instance and its functions to calculate ship propulsion power. Thus, Route provides functions to calculates routes, subroutes and the ship propulsion power for each route. The function calculateLEDM() calculates the Local Energy Distribution moments for the route.*

<<Java Class>>
⊖ **Route**
models.geography

▫ ship: Ship
▫ waypoints: Waypoint
▫ destination: Position
▫ step_distance: double
△ heuristic: HeuristicI

⊚ᶜ Route(Ship,Waypoint,Position,HeuristicI)
⊚ calculate_LEDM(Type):Vector<Double>
⊚ calculate_power(ExecutionMode):double
⊚ get_power():double
⊚ get_power_curve():Vector<Double>
⊚ calculate_waypoints(ExecutionMode):Waypoint
⊚ calculate_next_waypoint(Waypoint,double,ExecutionMode):Waypoint
⊚ get_mid(Waypoint,Position,ExecutionMode):Step
⊚ calculate_power_for_orthodrome(Waypoint,Position,ExecutionMode):double
⊚ get_basins():HashSet<Basin>
⊚ get_travel_time():double
⊚ from(Waypoint):Route
⊚ till(Waypoint):Route
⊚ sub_route(Waypoint,Position):Route
⊚ reverse():Route
⊚ scale(int):void

118 C. UML diagrams of the Java implementation

Figure C.2: The central Java class in the simulation is **Route** which is composed of a linked list of waypoints. Each waypoint is defined by its position, date and forecast horizon and the distance between two waypoints is calculated by the Orthodrome. Orthodromes are calculated based on a reference ellipsoid with a radius of 6.378 km and a flattening of $\frac{1}{298}$ (in accordance with the World Geodetic System 1984). A route contains an instance of class **Ship** to simulate the ship propulsion power. A route instance also provides its associated basins (i.e., oceans) and a list of Teleconnetction indices for these basins. Currently, only the NAOI is implemented for the North Atlantic basin.

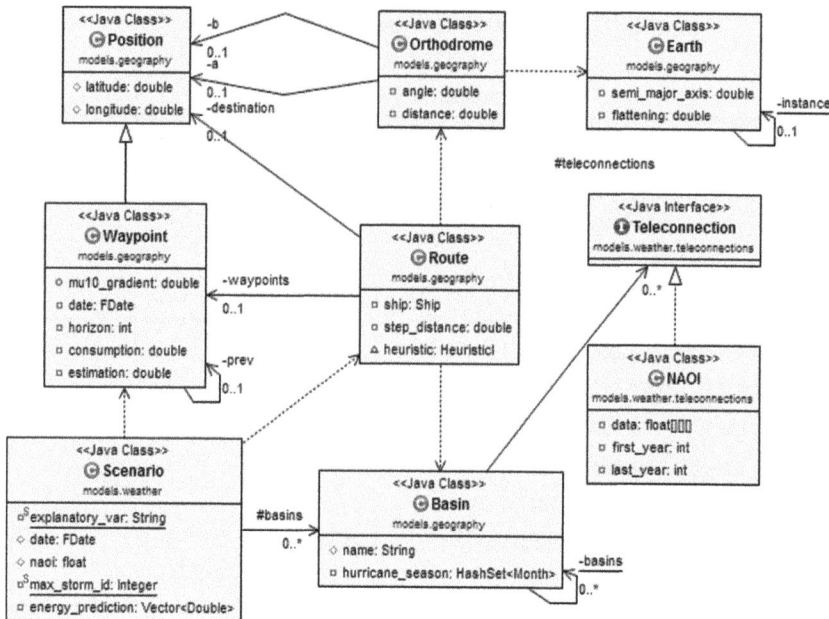

C. UML diagrams of the Java implementation   119

Figure C.3: The ship model is initialized with the basic ship properties. It uses objects of three classes to calculate wind- and wave resistances and the engine propulsion power saved by WaSP. Three wind drives are available: Flettner Rotor, k and the DynaRig, see chap. 6.

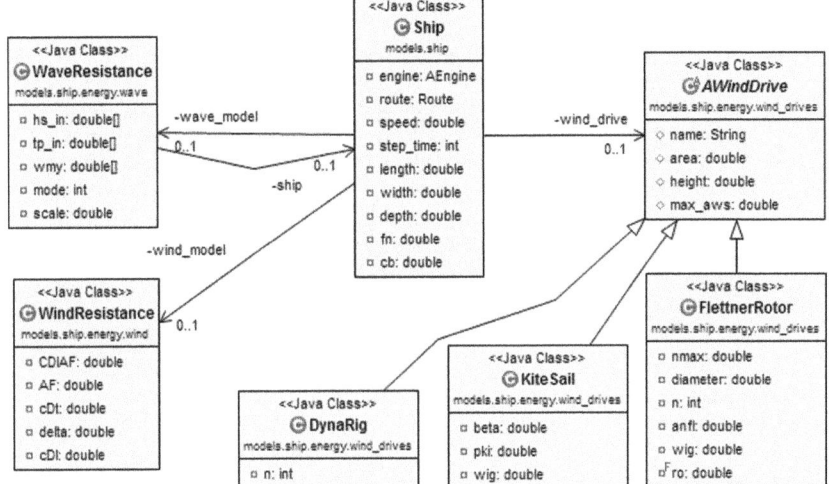

Figure C.4: h(x) *is implemented in the* `Forecast` *class. Other heuristics are also available. The class* `Minimum` *simply assumes a fixed value* e_{const} *for each sea mile from an arbitrary waypoint to the destination. Obviously,* e_{const} *depends on the type of ship and on the ship speed. If the class* `Distance` *is used as heuristic the A* search will rank waypoints according to their distance from the destination. Hence, the final route will be an orthodrome (shortest path) between departure and destination. The* `Average` *value heuristic is used to estimate the ship propulsion energy for routes that exceed the maximum forecast horizon of 168 hrs. Here the average ship propulsion energy from the first 168 hrs of the route is extrapolated to the ship propulsion energy for the remainder of the route. Finally, the* `Interval` *heuristic calculates a prediction interval for the ship propulsion energy from a waypoint to the destination and then the lower bound* E_{Min} *of the interval* ΔE *is returned by* h(x). *The class* `PredictionInterval` *internally calls the MATLAB function described in (Brabanter et al., 2013a) that implements the Least-squares Support Vector Regression (LSSVR).*

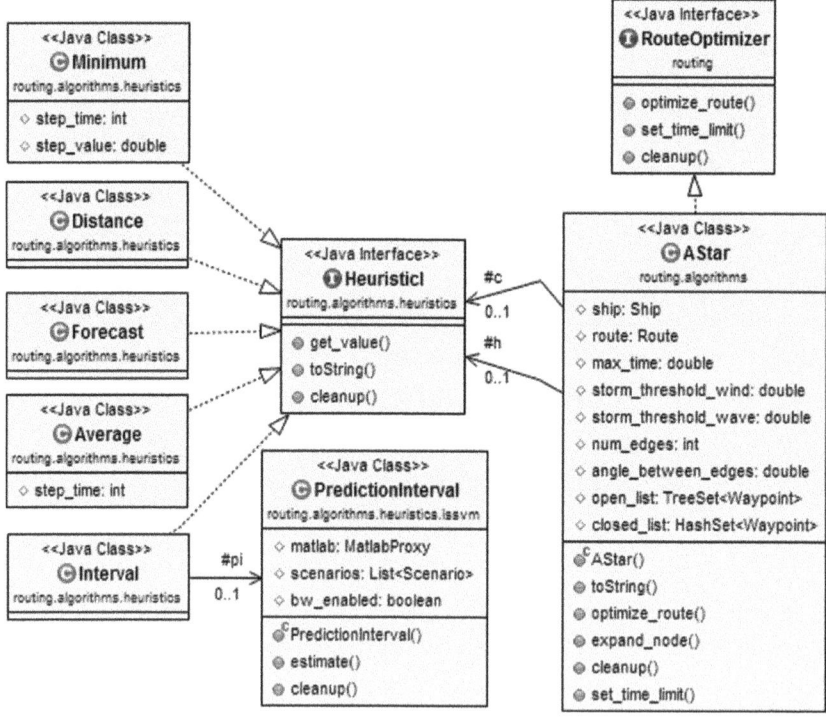

C. UML diagrams of the Java implementation

Figure C.5: UML sequence diagram showing the Java code that is involved in the calculation of prediction intervals. First the route between the waypoint and the destination is simulated with historic weather data (`simulate_historic_scenarios()`). Each scenario contains the predicted ship propulsion energy E_P and the true value E_A and several predictors which are described in Chap. 5. Then the scenarios are used to calibrate the quantile regression (QR) model (`train_LSSVR()`). This is done by a GNU R function call. The last step is another GNU R function call to the QR model with E_P and the predictors for the current scenario to calculate a prediction interval I^α (`run_LSSVR`). The `get_value()` function finally returns E_{Min} of I^α. This heuristic actually is admissible with probability $100 \cdot (1 - \alpha)$.

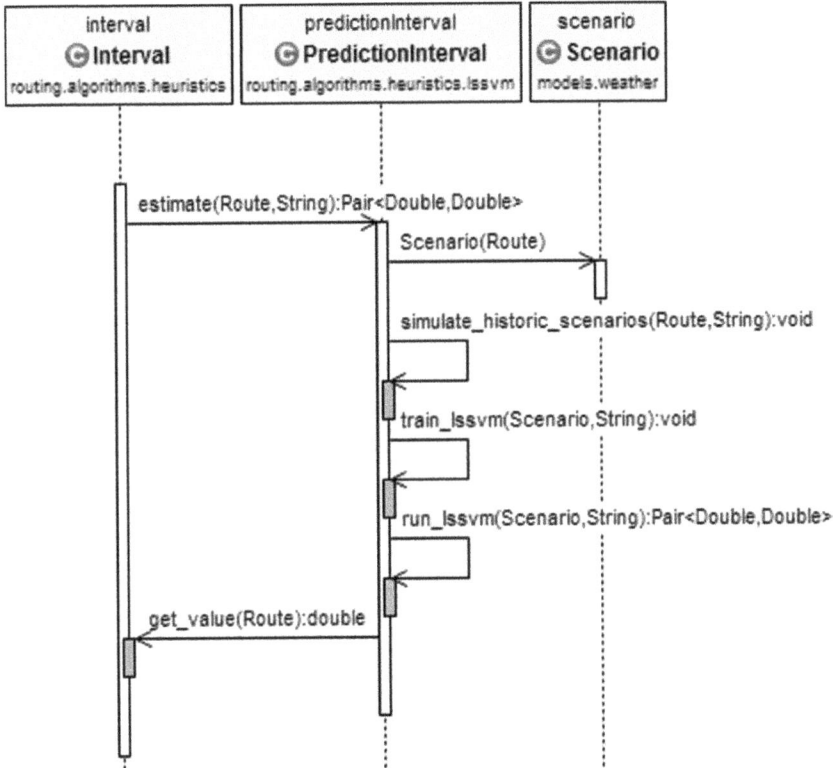

Figure C.6: UML class diagram of Java interface to MySQL database with archived historical weather analyses and predictions by the German Weather Service. The class `WeatherDBI` is an abstract interface that is implemented to integrate data from a numerical weather prediction (NWP) model. The class `Route` contains the method `calculate_LEDM()` to calculate the Local Energy Distribution Moments (sec. 5.2.5.1) as regressors for the quantile regression model.

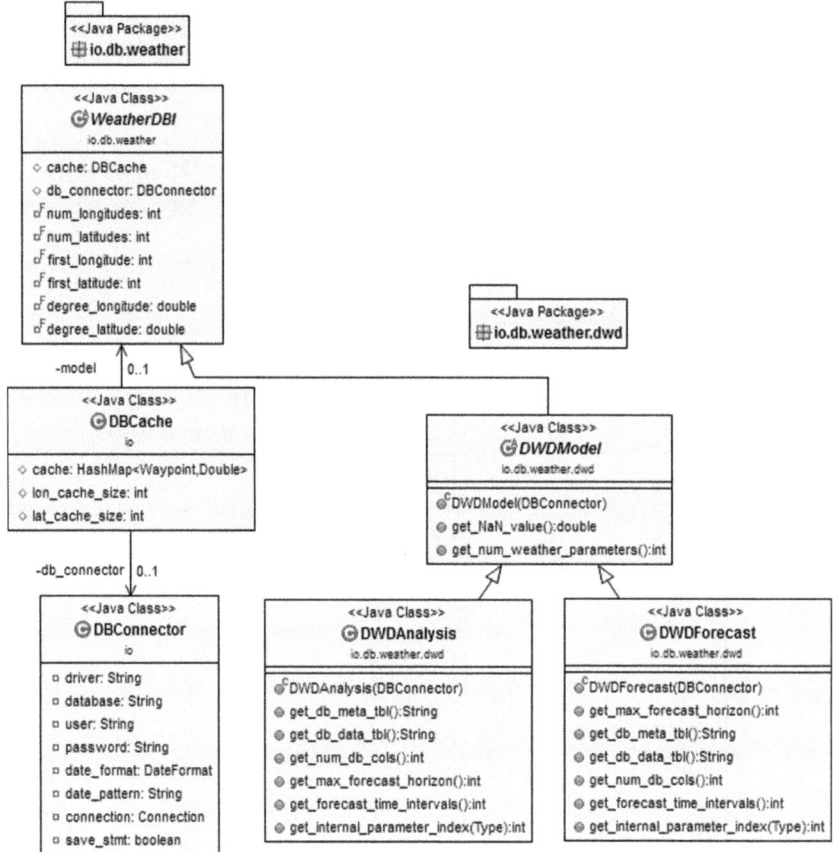

D. The Global Sea Model (GSM)

Table D.1: Outputs of the GSM model by the German Weather Service (DWD) as provided for this work.

abbreviation	description	units
MU10	mean wind speed 10 m above sea level	[m/s]
MDWI	mean wind speed 10 m above sea level	[°]
SWH	significant wave height	[m]
SHWW	significant height of wind waves	[m]
MDWW	mean direction of wind waves	[°]
MPWW	mean period of wind waves	[s]
SHPS	significant height of primary swell	[m]
MDPS	mean direction of primary swell	[°]
MPPS	mean period of primary swell	[s]

Table D.2: Time schedule for updates of weather analyses and weather predictions of the GSM model by the German Weather Service (DWD). t+00 UTC denotes weather analyses.

weather analyses (2005–February 2008):	t+03 UTC - t+06 UTC - t+09 UTC - t+12 UTC
weather analyses (February 2008–2012):	t+00 UTC - t+03 UTC - t+06 UTC - t+09 UTC
weather predictions:	t+00 UTC - t+03 UTC - t+06 UTC… (updated every 12 hours)

Maritime Logistik / Maritime Logistics

Herausgegeben von Prof. Dr. Burkhard Lemper und Prof. Dr. Frank Arendt

Ziel dieser Schriftenreihe des *Instituts für Seeverkehrswirtschaft und Logistik (ISL)* ist es, Aspekte und Entwicklungen aus den verschiedenen Bereichen der maritimen Logistikbranchen aufzugreifen und aktuelle Trends und Perspektiven zu diskutieren. Dabei meint Maritime Logistik – Maritime Logistics nicht nur die klassischen Bereiche wie Schifffahrt, Häfen, Schiffbau oder Verkehrspolitik, sondern schließt auch die Betrachtung der vielen weiteren Akteure in den globalen Transportketten ein, die direkt oder indirekt an den Logistikprozessen der maritimen Wirtschaft beteiligt sind.

Aufgegriffen werden logistische Fragestellungen zu Themen wie z.B. Hinterlandverkehr und intermodale Verkehre, Mesologistik und regionale Netzwerke wie GVZ und Logistikzentren, nachhaltige Geschäftsmodelle und Ressourceneffizienz oder Supply Chain Controlling. Im Mittelpunkt der Reihe *Maritime Logistik – Maritime Logistics* stehen aber auch informationslogistische Themen wie die Planung und Überwachung intermodaler Transportketten durch ein aktives Supply Chain Event Management, die Sicherheit und Transparenz im internationalen Containerverkehr oder die Planungsunterstützung und Optimierung logistischer Prozesse in Häfen und Terminals mit Hilfe quantitativer Methoden.

Band 1   Burkhard Lemper / Manfred Zachcial (eds.): Trends in Container Shipping. Proceedings of the ISL Maritime Conference 2008. 9th and 10th of December, World Trade Center Bremen. 2009.

Band 2   Hans-Dietrich Haasis / Holger Kramer / Burkhard Lemper (Hrsg.): Maritime Wirtschaft – Theorie, Empirie und Politik. Festschrift zum 65. Geburtstag von Manfred Zachcial. 2010.

Band 3   Kerstin Lange: Engpassorientierte Analyse der Ver- und Entsorgungslogistik von Steinkohlekraftwerken. Unter besonderer Beachtung der maritimen Logistik. 2011.

Band 4   Torben Möller: Finanzierung von Seehafeninfrastrukturen. 2012.

Band 5   Burkhard Lemper / Thomas Pawlik / Susanne Neumann (eds.): The Human Element in Container Shipping. 2012.

Band 6   Manuel Kühn / Karsten Seidel / Jochen Tholen / Günter Warsewa: Dryports - Local Solutions for Global Transport Challenges. A Study by the Institute Labour and Economy (IAW) of the University of Bremen. 2012.

Band 7   Philip M. Blumenthal: Optionsscheine auf Frachtraten. Modellierung und empirische Überprüfung für den Container-Seeverkehr. 2013.

Band 8   Iven Krämer: Die deutschen Seehäfen im Fokus überregionaler Entwicklungspolitik. Eine Folgenabschätzung zum Nationalen Hafenkonzept. 2015.

Band 9   David Zastrau: Estimation of Uncertainty of Wind Energy Predictions. With Application to Weather Routing and Wind Power Generation. 2017.

www.peterlang.com

www.ingramcontent.com/pod-product-compliance
Ingram Content Group UK Ltd.
Pitfield, Milton Keynes, MK11 3LW, UK
UKHW022154230426
12049UKWH00004BA/87